AUSTRALIAN NATURAL HISTORY SERIES

FUR SEALS AND SEA LIONS

ROGER KIRKWOOD AND SIMON GOLDSWORTHY

CSIRO

PUBLISHING

National Library of Australia Cataloguing-in-Publication entry

Kirkwood, Roger.

Fur seals and sea lions/by Roger Kirkwood and Simon Goldsworthy.

9780643096929 (pbk.)
9780643109834 (epdf)
9780643109841 (epub)

Australian natural history series.

Includes bibliographical references and index.

Australian sea lion – Australia.
Southern fur seals – Australia.
Sea lions – Habitat – Conservation.
Seals (Animals) – Habitat – Conservation.
Marine ecology – Australia.

Goldsworthy, S. D.

599.79750994

Published by
CSIRO PUBLISHING
36 Gardiner Road, Clayton VIC 3168
Private Bag 10, Clayton South VIC 3169
Australia

Telephone: [+613] 9545 8555
Local call: 1300 788 000 (Australia only)
Fax: +61 3 9662 7555
Email: csiropublishing@csiro.au
Web site: www.publishing.csiro.au

Front cover: Australian fur seal. Photographer: Vincent Anthony.

Back cover images by (clockwise from top left): Roger Kirkwood; Jay Town, *Herald Sun Melbourne*; Roger Kirkwood; Simon Goldsworthy; and Simon Goldsworthy

Set in 10.5/14 Adobe Palatino, Optima and Stone Sans
Edited by Joy Window
Cover and text design by James Kelly
Typeset by Desktop Concepts Pty Ltd, Melbourne
Printed by Ingram Lightning Source

CONTENTS

ACKNOWLEDGEMENTS

We thank Dr Peter Shaughnessy for his review of a draft and Dr David Slip for his review of the proofs. Extra photography was kindly supplied by Vincent Antony, Dr John Gibbens, Dr Michael Lynch, Tony Mitchell, Dr Richard Campbell, Dr Rachael Gray, Dr John Kirkwood, Jay Town, James Archibald, David Casper, Roy Hunt, David Hocking, Neville Johnson, Brad Page, Heidi Ahonen, Fran Solly, Warren Reed, Julia Back and Ken Mankey. Our respective organisations, Phillip Island Nature Parks and South Australian Research and Development Institute, supported our involvement with the book. We especially thank our wives (Marjolein, Jo) and kids (Jay, Emily, Max and Theo), who tolerated the restriction on our spare time. We greatly appreciated the editorial advice of Joy Window, and publication assistance from John Manger and Tracey Millen.

1

INTRODUCTION

Historical context

On 8 February 1797, the cargo ship *Sydney Cove* was bound for Sydney from Calcutta with a speculative cargo that included 7000 gallons of rum. Storms *en route* caused the ship (and maybe some rum) to leak badly and so, coming up the east coast of Tasmania, it was forced to beach beside an uncharted island, now called Preservation Island, in the Furneaux Group of south-eastern Bass Strait. All crew got ashore safely and much of the cargo was secured. After 2 months of recovery and planning, a ships' long-boat with 17 sailors headed north to seek rescue but after just a few days it was wrecked on the Victorian coast near Cape Howe. There followed an epic overland adventure from which three survivors reached Sydney.

A rescue mission was mounted. On its return to Sydney, survivors of the *Sydney Cove* and crew of the rescue vessel reported that islands in the Furneaux Group abounded with fur seals. Captain Charles Bishop of the brig *Nautilus* was passing through Sydney at the time and, on spec, decided to head to the Furneaux Islands for some sealing. He sailed from Sydney in October 1798 and returned in December with 5000 fur seal skins and 180 large 'hair' seal skins (sea lions). The *Nautilus* again sailed to the Furneaux Islands in January 1799 and returned to Sydney in March with a further 3800 skins, mostly fur seal. Bishop recorded that he had found '… the fur seal of the best quality in such numbers that we could average 200 skins a day'. There had been earlier sealing ventures in the south Pacific, including one that harvested 4500 New Zealand fur seal skins from Dusky Sound (NZ) in 1792. But these had not stimulated further immediate interest. In contrast, a 'rush' of sealers followed the *Nautilus* to Bass Strait, initiating an important industry for the fledgling colony of Australia. Over the next 20 years, the colony 'rode on the seal's back'. Bass Strait was quickly opened to sealing (Figure 1.1).

Between 1805 and 1808, 40–50 vessels with shore-based gangs were locating and harvesting seals in Bass Strait. In Sydney and Hobart, thriving ship building,

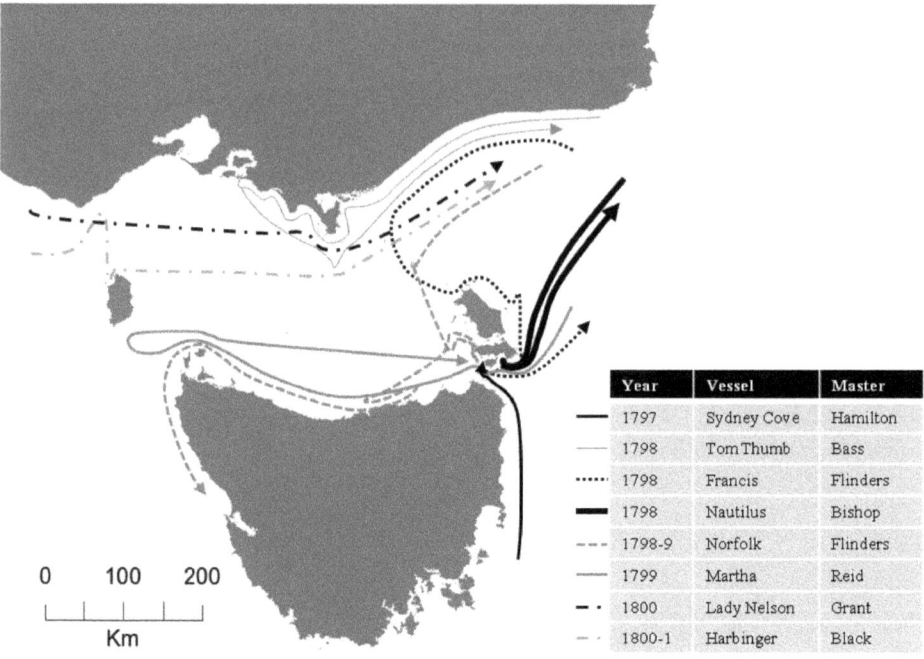

Year	Vessel	Master
—— 1797	Sydney Cove	Hamilton
—— 1798	Tom Thumb	Bass
······ 1798	Francis	Flinders
▬▬ 1798	Nautilus	Bishop
- - - 1798-9	Norfolk	Flinders
—— 1799	Martha	Reid
—·— 1800	Lady Nelson	Grant
—·— 1800-1	Harbinger	Black

Figure 1.1. Name, master and routes of vessels visiting Bass Strait between 1797 and 1800.

chandlery, crewing and marketing enterprises developed. Seal numbers diminished rapidly and by 1810 shore-based gangs ceased to operate in Bass Strait. The focus for sealers shifted to Kangaroo Island in South Australia, where a further 100 000 skins were obtained, and locations in New Zealand and the subantarctic. The Auckland Islands were discovered by sealers in 1806, and Campbell and Macquarie islands in 1810. By the 1830s, almost 1.4 million fur seal, 4000 Australian sea lion and 5600 New Zealand sea lion skins had been taken from the region. Elephant seals were eliminated from Bass Strait (a population of around 17 000 centred at King Island) and heavily harvested from subantarctic islands, particularly Macquarie (~200 000 seals).

The spectacular rise and demise of sealing in southern Australia mirrored a scenario that played-out worldwide between the 1770s and 1850s. Hundreds of vessels sought and exploited seal populations. Fur seal pelts were valued for both their leather and fur, which was removed and compressed into a felt for clothing. The principal market was China, where over 5 million fur seal pelts were sent. Significant outcomes of the industry included the establishment of economically sustainable colonies in Australia and elsewhere, and discoveries of numerous islands and stretches of coastline, including in Antarctica. In addition, the industry removed in a short period huge numbers of high trophic-level predators (those at the top of the food web) from many of the world's oceans, altering marine ecosystems prior to there being any substantial understanding of their structure and complexity. Most seal stocks are still to recover former ranges and abundances, and ecosystems are still rebounding.

Oceanography of southern Australia

Between 200 and 40 million years ago (mya), the southern supercontinent of Gondwana was splintering. Large land masses moved north, leaving Antarctica to settle alone over the South Geographic Pole. A circumpolar Southern Ocean isolated Antarctica from warmer northern waters and established oceanographic conditions that strongly affect marine productivity and climate in the Southern Hemisphere.

The continent of Australia is surrounded by circulating ocean currents (Plate 1a). Virtually all waters between 30° and 60° latitude in the Southern Hemisphere take up the dominant west to east flow of the Southern Ocean. This flow directs temperate and subantarctic surface waters over the southern coastline of Australia. Warm currents also feed into the region, via the Leeuwin Current down the Western Australian coast and east across the Great Australian Bight, and the East Australia Current down the east coast before peeling off into the Tasman Sea between 37° and 42° south. Sea surface temperatures across southern Australia range from 10 to 25°C and are warming. Large warm-core and cold-core eddy systems can develop off the east coast of Australia due to interaction between the warm East Australia Current and colder subantarctic waters. The oceans are warming relatively quickly off Australia's south-east coast. From 1900 to 1999, mean sea temperature increases off south-western (2.0°C) and south-eastern (2.3°C) Australia were considerably more than the global average (0.6°C). Ocean current strengths and sea temperatures across Southern Australia vary seasonally and inter-annually. Many sites occupied by seals in Australia are low-lying (Plate 1b) and will be inundated frequently over the next 100 years, particularly so with increases in sea level and storm frequencies that will result from climate change. This impact, along with marine ecosystem changes, will alter the habitat use and ranges of the seals.

Around southern Australia extends a 20 to 200 km wide continental shelf. The shelf is bounded by water depths of up to 5000 m, which represents a barrier against migration of shelf-dwelling species to and from southern Australian. The nearest temperate shelf waters are New Zealand, 2000 km to the east, and southern Africa, 8000 km to the west.

Deep off the southern Australian shelf resides cold Antarctic surface waters. These arrive after subduction at the mid-ocean Antarctic Polar Front (situated between 54° and 58° south) and can well-up onto the shelf. One mechanism for such upwelling is persistent south-easterly winds across south-eastern Australia. Typically, such winds occur between November and May, particularly when there is a 'blocking' high atmospheric pressure system over the Tasman Sea. The cold water may remain at depth on the shelf, resulting in a pronounced thermocline (temperature gradient) in the water column, or be drawn to the surface. The most familiar surface upwelling is on the Bonney Coast of eastern South Australia/western Victoria. Others occur south and west of Eyre Peninsula, off Kangaroo Island, and on the northern-west coast of Tasmania. In winter months, the eastward flowing Leeuwin Current dominates along the southern coast of Australia. This current carries warm, saline water into the Great Australian Bight, where it enhances down-welling and mixing of the water column. Such activity in winter, coupled with the low irradiance and short day lengths, results in low productivity.

On the inner shelf off South Australia there is a westerly moving, boundary current, termed the Flinders Current System. This has oceanographic, biological and ecological similarities to the eastern boundary current systems off the west coasts of Africa (Benguela Current) and North America (California Current). Levels of primary production and fish production within the Flinders Current System are higher than those in other parts of Australia.

Variability in ocean current strength is strongly influenced by winds which change within and between years. Across southern Australia, the most persistent strong winds are the winter westerlies, so westerly currents are strongest in winter. Inter-annual variability in ocean current strength is influenced by broad-scale processes, including the El Niño Southern Oscillation (ENSO) and the Southern Annular Mode (SAM). The ENSO acts in an enigmatic fashion, oscillating between what are termed El Niño and La Niña states in a pattern that has proven difficult to forecast. Reduced atmospheric pressure in the Central-Eastern Pacific Region initiates the onset of El Niños, which promote drought conditions in Australia. One product of the reduced pressure in the Australian region is reduced strength of the Leeuwin and East Australian Currents. Wintertime circulation along Australia's southern shelf region is reduced, but without a compensatory increase in summer currents. The reduced circulation appears to enhance upwelling along the southern Australian coast.

In La Niña states, which provide wetter years in Australia, the warm East Australian Current is enhanced and extends further down the shelf to the east coast of Tasmania. This reduces the extent of subantarctic waters and may stem westward flow through Bass Strait. The Leeuwin Current is also enhanced by La Niña conditions.

Variations in the SAM are the result of the coupling of atmosphere and sea conditions that pulse weak colder-than-average, weak warmer-than-average, strong colder-than-average, then strong warmer-than-average conditions in a northerly direction on an 8–12-year cycle. The fluctuations influence multi-year cycling in the strength of zonal (westerly) and meridional (southerly) winds, which Ron Thresher, from CSIRO Fisheries, and others have linked to variability in fish recruitment and even strandings of cetaceans in south-east Australia.

In 2010, an Australia-wide project, the Integrated Marine Observing System (IMOS), was established to coordinate data gathering, utilising research and commercial vessels, remote recording devices and even devices on seals to monitor currents around Australia. Ongoing research by John Middleton, from the South Australian Research and Development Institute, and others is improving our knowledge, but there is a lot more to find out about the strengths, directions, variability and effects of the water bodies around southern Australia.

Regional habitats

The largely transitional, cool temperate seas around southern Australia are referred to as the Flindersian Biogeographic Region. Within this are broad provinces distinguished by factors that influence its productivity, such as prevailing current, aspect, temperature and breadth of the continental shelf. The provinces are separated by zones of overlap termed biotones. The South-West Province around south-western Australia has a narrow shelf affected by the Leeuwin Current. A broadening of the shelf extends a biotone across the south-facing Great Australian Bight to the Gulf Province of South Australia. Waters there can be highly productive, being influenced by the permanent offshore presence and occasional upwelling of nutrient-rich Antarctic surface water and the Flinders Current.

The Bass Strait Province between mainland Australia and Tasmania constitutes a shallow (less than 100 m deep) marine basin with low productivity and slow flow rates: it can take up to 160 days for surface water to cross the strait. Through-water movement is mostly wind-driven from west to east but is pulsed by tidal currents and back eddies. A province around southern Tasmania (sometimes referenced separately as the Maugean Region) comprises colder waters. Up the east coast of Australia is the Central Eastern Province, with the distinctly warmer waters of the East Australia Current.

Scope of this book

In this book, we focus on the three seals that currently breed and live in coastal waters adjacent to continental Australia: the Australian fur seal (*Arctocephalus pusillus doriferus*), the New Zealand fur seal (*Arctophoca forsteri*, also known as *Arctophoca australis forsteri*) and the Australian sea lion (*Neophoca cinerea*). We also present information on species that do, and in the near future could, visit continental Australian waters. These include seals that breed in the Antarctic sea-ice zone, and seals that breed at subantarctic islands of the South Pacific and Indian Oceans.

2

EVOLUTION AND RECENT HISTORY

Evolution is a continual process of genetic mutation, hybridisation and adaptation. Species change, new species emerge and others go extinct. One means for speciation is geographical isolation. Through unique habitat stimuli and advantages or disadvantages afforded to genetic mutations, separated populations can progressively become distinct. There may be a simple split into two identifiable groups or a 'cline' (gradation), in which populations at either end of the range differ most and there is a gradual transition of characteristics across intervening groups. A second means for new species to emerge is through hybridisation (interbreeding) between similar species, very occasionally producing viable and fertile young. Gradually, through generations, one species may take on attributes of another, or a new species could emerge with attributes of both parent taxa.

'What constitutes a distinct species?', 'how closely related are species that have similar characteristics?' and 'how long ago did species emerge?' are hard questions. Taxonomists often have different opinions about what constitutes a species. Some become 'splitters' and readily separate populations into unique species, while others are 'lumpers' and tend to combine like-populations into single taxon. The progress continually changes the field of taxonomic research. Originally, taxonomists distinguished species primarily on morphological characteristics. Now, genetic material in combination with morphology, behaviour and distribution are used. Rapid communication through the scientific literature greatly assists the dissemination of ideas now, and contrasts hugely with the levels of communication available to taxonomists even 20 years ago. And for taxonomists, the past is also subjected to change. Ongoing archaeological discoveries and analysis techniques modify recognitions of species links and dates of emergence.

Evolution of seals

The pinnipeds (seals) are a monophyletic group (derived from a single evolutionary lineage) of aquatic carnivores that is most closely related to either mustelids (otters) or ursids (bears). They originated during the Oligocene Epoch (38 to 24 mya, Figure 2.1). Following emergence, there was a basal split around 33 mya between lines that evolved into the three current families: Otariidae (eared seals), Phocidae (earless or true seals) and Odobenidae (walrus). Species of pinnipeds have come and gone. Currently, 35 extant species are recognised: 16 otariids, 18 phocids and one odobenid, *Odobenus rosmarus*.

Evolutionary and taxonomic links between animals are constantly being updated as we learn more about species and their evolution, discover new fossils, and derive new techniques for analyses. As a result, there can be disagreement between different taxonomists and between analysis methods (an example of relatedness between extant species is presented in Figure 2.2). Taxonomic relationships between some pinniped species remain inconclusive, particularly between species that derived from recent and rapid radiations (like the fur seals). Here we present a current state of play in nomenclature and phylogeny summarised from literature up to 2012. At the time there was no complete consensus and some relationships described here are likely be interpreted differently in the future. Key contributors to this field of research include Judith King, Louise Wynen, Sylvia Brunner, Jeff Higdon, Takahiro Yonezawa, Annalisa Berta and Morgan Churchill.

Otariids

Modern otariids are thought to have evolved from the ancestral family Enaliarctidae in the Northern Pacific region circa 11 to 8 mya (Figure 2.3). Traditionally, otariids have been separated into the subfamilies Arctocephalinae (fur seals) and Otariinae (sea lions) based on the presence or absence of underfur. However, there is molecular and morphological evidence to suggest a more complicated evolutionary history.

The northern fur seal *Callorhinus ursinus* appears to have been derived from the ancestral Enaliarctidae and has remained in the Northern Pacific. It could be a sister

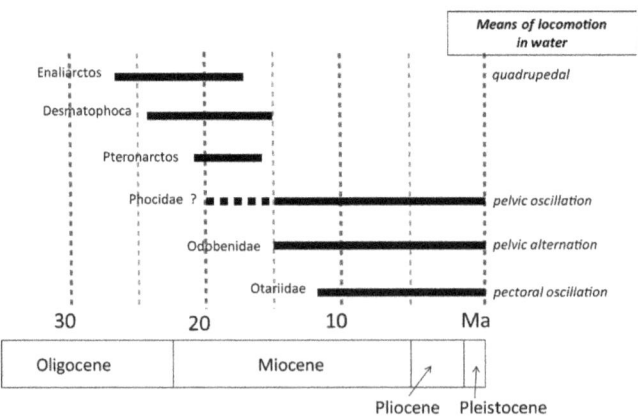

Figure 2.1. Chronological ranges of extinct and extant pinnipeds (Ma = million years ago).
Adapted from Berta and Sumich (1999).

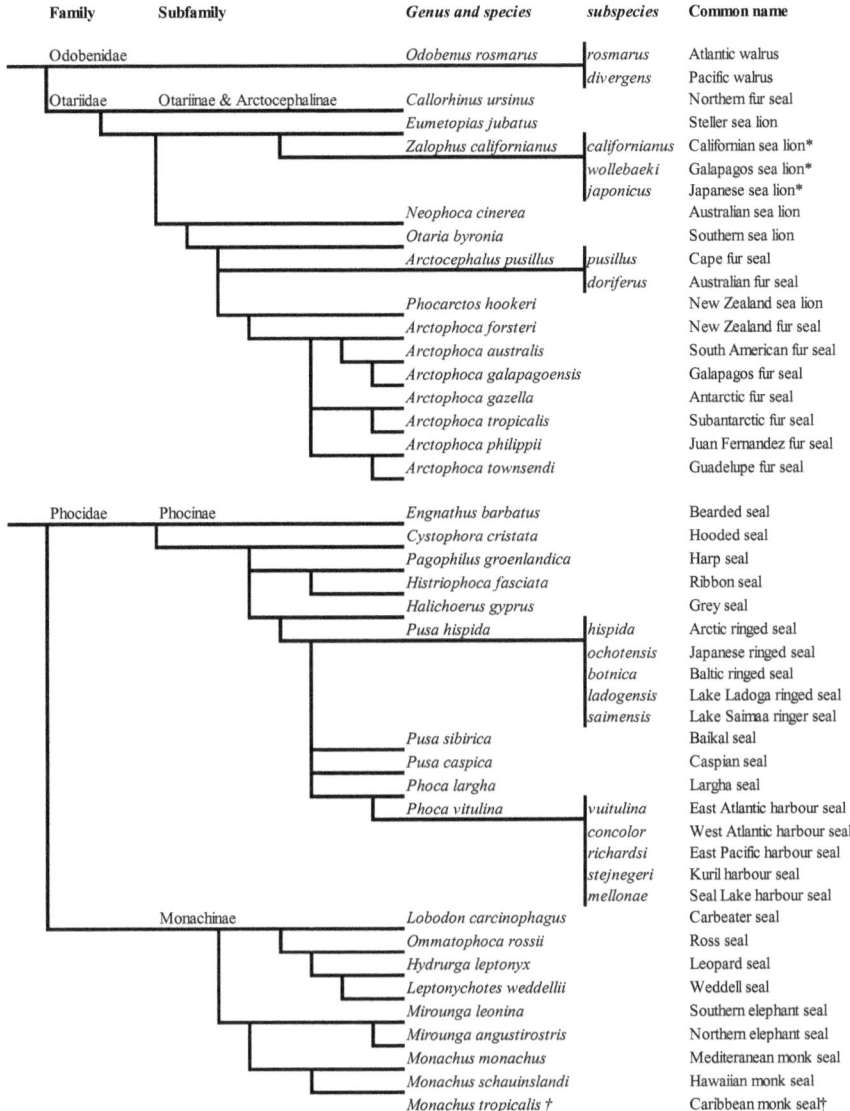

Family	Subfamily		Genus and species	subspecies	Common name
Odobenidae			Odobenus rosmarus	rosmarus	Atlantic walrus
				divergens	Pacific walrus
Otariidae	Otariinae & Arctocephalinae		Callorhinus ursinus		Northern fur seal
			Eumetopias jubatus		Steller sea lion
			Zalophus californianus	californianus	Californian sea lion*
				wollebaeki	Galapagos sea lion*
				japonicus	Japanese sea lion*
			Neophoca cinerea		Australian sea lion
			Otaria byronia		Southern sea lion
			Arctocephalus pusillus	pusillus	Cape fur seal
				doriferus	Australian fur seal
			Phocarctos hookeri		New Zealand sea lion
			Arctophoca forsteri		New Zealand fur seal
			Arctophoca australis		South American fur seal
			Arctophoca galapagoensis		Galapagos fur seal
			Arctophoca gazella		Antarctic fur seal
			Arctophoca tropicalis		Subantarctic fur seal
			Arctophoca philippii		Juan Fernandez fur seal
			Arctophoca townsendi		Guadelupe fur seal
Phocidae	Phocinae		Engnathus barbatus		Bearded seal
			Cystophora cristata		Hooded seal
			Pagophilus groenlandica		Harp seal
			Histriophoca fasciata		Ribbon seal
			Halichoerus gyprus		Grey seal
			Pusa hispida	hispida	Arctic ringed seal
				ochotensis	Japanese ringed seal
				botnica	Baltic ringed seal
				ladogensis	Lake Ladoga ringed seal
				saimensis	Lake Saimaa ringer seal
			Pusa sibirica		Baikal seal
			Pusa caspica		Caspian seal
			Phoca largha		Largha seal
			Phoca vitulina	vuitulina	East Atlantic harbour seal
				concolor	West Atlantic harbour seal
				richardsi	East Pacific harbour seal
				stejnegeri	Kuril harbour seal
				mellonae	Seal Lake harbour seal
	Monachinae		Lobodon carcinophagus		Carbeater seal
			Ommatophoca rossii		Ross seal
			Hydrurga leptonyx		Leopard seal
			Leptonychotes weddellii		Weddell seal
			Mirounga leonina		Southern elephant seal
			Mirounga angustirostris		Northern elephant seal
			Monachus monachus		Mediteranean monk seal
			Monachus schauinslandi		Hawaiian monk seal
			Monachus tropicalis †		Caribbean monk seal†

Figure 2.2. A molecular supertree of the world's extant pinnipeds (plus the recently extinct *Monachus tropicalis*) based on a weighted matrix representation with parsimony analysis of 50 maximum-likelihood gene trees. Branch lengths correspond to time with the scale bar indicating one million years. Boxed subset provides additional detail on branching order for two parts of the supertree where divergences occurred over a short timeframe. *Note that the three sea lions in the genus *Zalophus* are now considered to be distinct species, rather than subspecies. Derived from Higdon *et al.* (2007). © BioMed Central.

species to all extant otariids. Around 8 to 6 mya, there was an ancestral crossing of early otariids into the Southern Hemisphere. There, between 6.1 and 3.4 mya, genetic lines were established to the extant sea lion genera (*Neophoca*, *Eumetopias*, *Zalophus*, *Otaria* and *Phocarctos*) and to the fur seal *Arctocephalus pusillus* (which arose circa

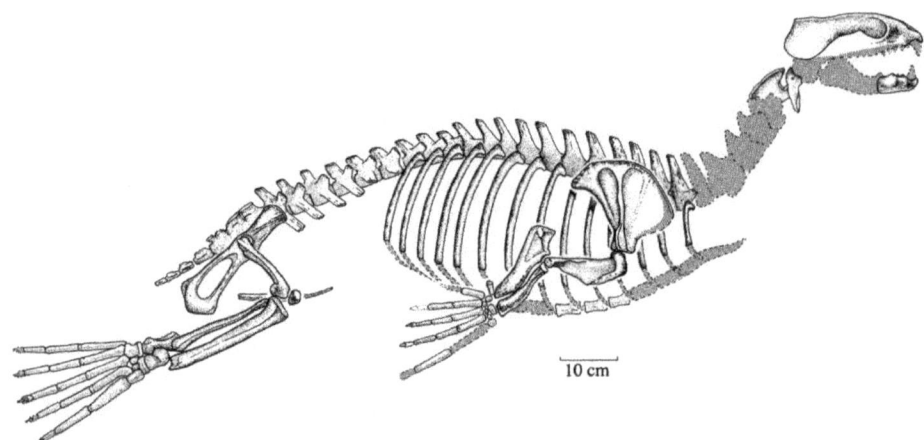

Figure 2.3. Skeletal reconstruction of *Enaliarctos mealsi*. From Berta and Ray (1990), reproduced with permission of Taylor & Francis Group, LLC.

4.3 mya). The Australian sea lion, *Neophoca cinerea*, is the only extant species within its genus but an extinct species, *N. palatina*, has been recognised in fossils found in New Zealand.

The fur seal *Arctocephalus pusillus* (Cape and Australian fur seals) shows a commonality to several of the sea lions. For instance, its nearest extant relative appears to be the southern sea lion *Otaria byronia*. Apart from *A. pusillus*, all other extant Southern Hemisphere fur seals probably derived from a rapid radiation largely within the past million years. Up until 2011, these fur seals had been included in the genus *Arctocephalus*. In 2011, Annalisa Berta (San Diego State University) and Morgan Churchill resurrected for them the genus name *Arctophoca*. This separation within the southern fur seals made a lot of sense, based on the species's evolutionary history, morphology, behaviour and genetics, and had been mooted for some time.

Phocids

Prior to the middle Miocene Epoch, circa 16 mya, the family Phocidae had split into two subfamilies: the Monachinae largely in the Southern Hemisphere (eight species), and the Phocinae (10 species), which inhabit the Arctic and subarctic regions. Within the Monachinae, there are three lineages: a lineage leading to monk seals (*Monachus*) which separated circa 11 mya; a lineage leading to the elephant seals (*Mirounga*) which split circa 10 mya; and a lineage leading to lobodontine (Antarctic) seals. Amongst the lobodontine seals, crabeaters (*Lobodon*) diverged 7.1 mya, Ross (*Ommatophoca*) at 6.8 mya, and leopards (*Hydrurga*) and Weddells (*Leptonychotes*) at 4.3 mya. The two elephant seals (northern and southern) split from one another between 2 and 1 mya.

Pinniped presence in Australian waters

Pinnipeds established in the Southern Hemisphere tens of millions of years after the separation from Antarctica of the Australian and South American continental plates around 40 mya. Phocids established in tropical, temperate and Antarctic waters while initial radiation of the otariids was into temperate regions. Evolutionary radiation during the past one million years of the thick-furred southern fur seals possibly coincided with colonisation of colder subantarctic and Antarctic waters.

Australian and New Zealand southern coastal waters were first colonised by ancestors of Australian and New Zealand sea lions between 6 and 4 mya. Elephant seals may have colonised earlier, although their presence around continental Australia might also have been a recent expansion from a more southerly range. Ancestral New Zealand fur seals established populations around both New Zealand and Australia in the last one million years. Finally, the *A. pusillus* genus, which evolved in southern African waters around 4 mya, arrived in Australian coastal waters, establishing the Australian fur seal population, between 18 000 and 12 000 years ago. This coincided with the end of the last ice age, when there was a sea level rise of approximately 100 m.

During the period of residence of seals in the Australasian region, sea levels have risen and fallen several times, successively submerging potential colony sites, then isolating them inland. The seals would have been forced to colonise new sites during each substantial sea level change.

Arrival of early human hunters in Australia about 80 000 years ago and in New Zealand about 1000 years ago undoubtedly influenced the distribution of breeding and haul-out sites of the seals. Those on easily accessible mainland coasts and nearby islands would have been subjected to hunting and a general shift to more isolated locations probably resulted.

The sealing era

The Dutch were the first Europeans to report seals in the Australian region. In 1629, the ship *Batavia* was wrecked on the Houtman Abrolhos and survivors ate the local sea lions. In 1642, Abel Tasman in the *Heemskerck* and *Zeehaen* recorded a seal at sea in the Great Australian Bight, and in 1727 survivors of the *Zeewyk* also ate sea lions at the Houtman Abrolhos. Then came the British. In 1769 and 1773, James Cook on *Endeavour* noted seals at sea while passing New Zealand, and in 1788 John Hunter of the *Sirius* (First Fleet) recorded seals at sea south of Tasmania. Four years later (in 1792) there was a harvest of seals from Dusky Bay, New Zealand. But sealing in the region really kicked off in 1798–99, when Charles Bishop harvested 9000 pelts from several months work in the Furneaux Group of eastern Bass Strait.

Details of the sealing industry in southern Australia are scarce. John Cumpston collated data on ship records and export transactions to provide valuable summaries of the industry in Bass Strait, Kangaroo Island and Macquarie Island. John Ling, following his retirement from the South Australian Museum, summarised much of the catch data from the sealing era (Table 2.1). In Bass Strait, harvests peaked between 1805 and 1808, and then declined due to resource depletion. They had largely ceased by 1810, although commercial takes were made into the 1820s and occasional takes to supply local markets continued through the next 100 years. From Bass Strait, sealers rapidly spread out, discovering and harvesting stocks across southern Australia, New Zealand, subantarctic islands and the Antarctic pack ice. John Ling collated historic records of cargoes of otariid skins and elephant seal oil from Australia and New Zealand in the 18th, 19th and 20th centuries. By 1840, at least 350 000 otariid pelts had

Table 2.1. Number of fur seal skins recorded in cargos taken from Australia and New Zealand, by region and decade. Data from Ling (1999, 2002).

Decade	Bass Strait	Kangaroo Is.	West Aust.	Macquarie Is.	NZ	Antipodes Is.	Bounty Is.	Auckland Is.	Campbell Is.	Total
1781–1790	.	.	.	.	.	.	.	.	.	.
1791–1800	15 600	.	.	.	25 108	.	.	.	.	40708
1801–1810	151 815	25 143	.	.	143 489	383 029	53 500	5000	15 200	777176
1811–1820	45 701	43 005	.	172 082	69 284	.	.	.	.	330072
1821–1830	26 782	26 477	7954	24 711	91 849	258	28	30 700	6194	214953
1831–1840	925	5 046	.	2471	19 266	.	.	4350	340	32398
1841–1850	.		.	.	147	.	.	.	.	147
1851–1860	300	.	.	.	.	.	.	.	.	300
1861–1870	.	.	.	.	.	.	.	.	.	.
1871–1880	.	.	.	60	11	.	620	550	125	1366
1881–1890	.	.	.	.	.	.	.	537	.	537
1891–1900	.	.	.	60	.	.	300	1160	516	2036
1901–1910	.	.	.	.	.	.	.	.	.	.
1911–1920	.	20	494	60	391	.	.	500	7093	8558
1921–1930	750	.	.	.	.	.	.	.	712	1462
1931–1940	.	.	.	.	.	.	.	.	102	102
1941–1950	691	.	.	.	7 668	.	.	.	.	8359
Total	242 564	99 691	8448	199 444	357 213	383 287	544 48	42 797	30 282	1 418 174

come from southern Australia, 320 000 from New Zealand, and almost 700 000 from the subantarctic islands of Macquarie, Campbell and Auckland.

Sealers undoubtedly could distinguish between the species of otariids they harvested. There was the 'haired seal' (Australian sea lion) and two fur seals, the abundant 'brown' (Australian) and a less abundant 'black' (New Zealand), which possessed the 'finest quality pelt'. The sealers seldom recorded this information though, nor did they record the exact locations of harvests and there are few images of seals from the sealing period (Figure 2.4). Thus, reconstructing species distributions of harvests is difficult. Bob Warneke attempted this, however, drawing from sealers records and accounts of explorers such as Matthew Flinders, whose observations date back to 1788–89.

Warneke recognised that Australian fur seals likely provided the majority of the estimated 244 000 skins exported from Bass Strait. He determined that Australian fur seals possibly pupped at over 26 sites in south eastern Australia prior to the commencement of sealing. Sealers also took New Zealand fur seals at four sites (at least) in Bass Strait, including the 1798 harvest of 9000 pelts at Cone Point on Cape Barren Island, in the Furneaux Group. Another colony was probably at Cat Island off the east coast of Flinders Island, with possible colonies in the Kent Group of eastern Bass Strait, and at Seal Rocks (also known as Saltpetre Rocks) on the west coast of King Island. West of Bass Strait, across the coasts of South and Western Australia, the New Zealand fur seal was more abundant and probably provided the majority of the 100 000 fur seal pelts from the vicinity of Kangaroo Island.

Australian sea lion harvests to 1840 included records of 4116 animals: 2110 from Kangaroo Island, 1521 from the Bass Strait islands (although 500 of these could have originated from Kangaroo Island) and 485 from WA, which includes 147 sea lions killed by survivors the ship-wrecked *Zeewyk*, on the Abrolhos Islands in 1727. Records indicate Australian sea lions once occurred throughout their current range from the Abrolhos Islands in the north-west to The Pages Islands in the south-east. Although there are no records of pupping by Australian sea lions in Bass Strait, considering the population size of some breeding colonies is less than 100 individuals and over 1000 were harvested in Bass Strait, it is probable that pupping did occur there. Likely sites for this were in the southern Furneaux Group (Clarke, Passage and Battery islands) and the Kent Group.

In addition to the otariids of southern Australia, a population of 10 000 to 17 000 southern elephant seals (*Mirounga leonina*) at King Island, Bass Strait, was eliminated by sealers.

Protection

Bob Warneke has collated information on the progressive protection of seals in Australian waters. In the late 1800s, long after the populations had been reduced to remnants of pre-harvest levels, the first regulations were introduced. These aimed to prevent local extinctions and rebuild the once profitable resource. New Zealand took the lead in 1875 by prohibiting sealing in its waters between October and May. In Australian states, initial regulations were in the form of closed seasons and a requirement for a 'permit to harvest'. These regulations came into effect sequentially in Tasmania (1889), Victoria (1891, via the Victorian *Game Act 1890*), Western Australia (1892), New

Figure 2.4. (a) *Seal Shooting in Bass's Straits*, 4 May 1881. **Artist unknown.** Source: La Trobe Picture Collection, State Library of Victoria, IAN04/05/81/92. **Published in** *Illustrated Sydney News*, 10 June 1882. The seals depicted are phocids, possibly Northern Hemisphere harbour seals, suggesting the artist had no personal experience of seals in Australian waters. (b) *The Australian sea bear or fur seal, Western Port*, by John James Wild, 2 November 1882. Drawn from specimens collected at Seal Rocks and published in McCoy (1885). Source: Museum Victoria, Image no. 000027208.

South Wales (1918) and South Australia (1919). The legislations were not policed and had little initial impact, as the open season was summer, when fur seals gather to breed and are most easy to harvest. The number of people undertaking sealing gradually decreased, as markets dried up and the arduous life style was avoided. Sealing in Bass Strait continued until 1923, when the open season shifted from summer to winter. The last sealers were operating from small wooden boats out of Cape Barren Island and the harsh conditions of winter made it too dangerous for them to navigate in Bass Strait and land on the remote sites.

Coastal and offshore fisheries developed in south-eastern Australia during the late 1800s and early 1900s. Fur seals were often viewed by the fishermen as competitors. For example, agitation by commercial fishermen in Victoria led to culls being authorised under tender in 1908, 1920, 1934, 1941 and 1948, although no tenders were received on three of the five occasions. In 1934, a ministerial permit to the Hastings Fisherman's Association allowed a cull of Australian fur seals at Seal Rocks. The cull was to be supervised by the then Chief Inspector of Fisheries and Game, Fred Lewis, but the fishermen informed the officer on the day of their departure and he could not travel from Melbourne in time. About 100 seals are reported to have been shot in 1 day. In 1948, culls proceeded at Seal Rocks and Lady Julia Percy Island under permits to fishermen, on condition that the carcasses were saved and shipped to Melbourne for commercial processing. A total of 691 carcasses were recovered and transported. The commercial processing failed because the abattoir equipment could not cope with the seals' body shape and the meat was considered to be unpalatable.

While agitating for further culls, fishermen shot at seals opportunistically, including at colonies and haul-outs. This practice was broadly thought of as legitimate. The opinion may have been based on the giving of a curious 'general authority' to fishermen of Western Port, Victoria, in 1929, which tolerated the killing of individual seals that were directly interfering with fishing gear. The 1929 edict for Western Port fishermen may have gradually morphed into an opinion across southern Australia that seals could be shot as required.

In 1975, all seals in Australian waters became protected, under the *National Parks and Wildlife Conservation Act 1975* (Cth) in Commonwealth waters and under various legislations in state waters. This protection, along with a gradual shift in societal attitudes toward seals (seeing them more as virtuous rather than nuisances) and closer management of fisheries procedures have undoubtedly reduced the rate of seal deaths due to interactions with fisheries, but some still continue.

Taxonomic recognition

In Australia today, we recognise three locally breeding species of otariid seals: the Australian sea lion, *Neophoca cinerea*: the Australian fur seal, *Arctocephalus pusillus doriferus*; and the New Zealand fur seal, *Arctophoca forsteri* (also known as *Arctophoca australis forsteri*). This has not always been the case, however (Table 2.2). In the late 1700 and early 1800s, sealers distinguished between 'haired seals' (sea lions) and 'furred seals'. They called male haired seals 'bulls', male furred seals 'wigs', all females 'klapmatches' and the young 'pups'. Some records from the early 1800s describe the presence of two types of fur seal in southern Australia: 'the black fur seal' with the 'finest quality fur', now assumed to be the New Zealand fur seal; and the 'brown', 'large' or

Table 2.2. Latin names given to the otariids of southern Australia.

Year	Australian sea lion	Author	New Zealand fur seal	Author	Australian fur seal	Author
1802	Otaria cinerea	Péron				
1802	O. albicollis	Péron				
1816	Neophoca cinerea	Péron				
1826	O. australis	Quoy & Gaimard			Otaria cinerea	Quoy & Gaimard
1828			Otaria forsteri	Lesson		
1828			Arctocephalus forsteri	Lesson		
1925			(species recognised	Le Souëf	Arctocephalus doriferus	Jones
1926			in Australia)		A. tasmanicus	Scott & Lord
1969					A. pusillus[1]	King
1971					A. pusillus doriferus	Repenning et al.
2011			Arctophoca forsteri[2]	Lesson		

[1] A. pusillus = Cape fur seal, named by Shreber in 1776
[2] Arctophoca resurrected by Berta and Churchill (2011)

'Tasmanian' fur seal or 'the fur seal of the southern seas', now assumed to be the Australian fur seal.

The first person to show a taxonomic interest in the Australian otariids was the French naturalist, Francois Péron, who traversed southern Australia with Captain Nicolas Baudin, from 1800 to 1804. In 1802, Péron in the ship *Le Naturaliste* (Baudin's command vessel was *Le Geographe*) arrived at Kangaroo Island. He saw mainly female and juvenile Australian sea lions there. In the Latin vernacular, he named them *Otaria cinerea* (*otaria* = little ear, *cinereus* = ash coloured). When he published his works in 1816, however, he chose the genus name *Neophoca* (new seal) rather than *Otaria*. On leaving Kangaroo Island, Péron travelled west to the Nuyts Archipelago where he encountered mainly male Australian sea lions, which possess a distinct white nape (back of neck). He unfortunately thought they were a different species and named them *O. albicollis* (white collared). This created confusion for later taxonomists and the confusion was intensified in 1826, when naturalists Jean Quoy and Joseph Gaimard, aboard *L'Astrolabe* on Dumont D'Urville's circumnavigation of the world, presumed the Australian fur seals they were seeing in Bass Strait were Péron's *O. cinerea*. They collected a male skull and skin from 'Westernport Bay' (probably Seal Rocks), Victoria, and lodged it in the Museum National d'Histoire Naturelle in Paris as 'Otarie cendrée, mâle. *Otaria cinerea*, Péron'. Venturing on to Western Australia, Quoy and Gaimard encountered actual Australian sea lions. Recognising these as different to the Bass Strait seals and thinking them also different to Péron's *O. albicollis*, they named them *O. australis* (southern). Thus, the Australian sea lion had three Latin names, with the 'assumed' type specimen ('type' indicates material, usually stored in a museum, on which the taxonomic description of a species is based) for one of these actually being an Australian fur seal.

It was not until 1969 that Judith King, then at the University of New South Wales, correctly identified the skull held at the Museum in Paris as being from a fur seal. King also confirmed *Neophoca cinerea* (Péron, 1816) as the legitimate Latin name for the only sea lion in Australia. Other common names attributed to this sea lion (often acknowledging the male's distinctive blonde nape) have been the 'white-necked hair seal', 'white capped sea lion', 'cowled seal', 'counsellor seal' and 'South Australian sea lion'.

The first person to apply a Latin name to either of the fur seals resident in Australia was René Lesson, the French naturalist on Louis Duperrey's 1828 *Coquille* expedition to the Pacific. Lesson called the fur seals of New Zealand *Otaria forsteri*. His species name acknowledged drawings of the seal by artist and naturalist Johann Georg Adam Forster of James Cook's 1772–75 expedition. The genus name *Otaria* did not stick for long because just prior to 1828 Georges Cuvier, at the Museum National d'Histoire Naturelle in Paris, had coined *Arctocephalus* (bear head) for all Southern Hemisphere fur seals. Hence, Lesson's seal became known as *Arctocephalus forsteri* (Lesson 1828) with the common names of the New Zealand, or Forster's, fur seal. This name was not formally applied to the 'black fur seal' of Australian waters, for a further 100 years.

Then, in 1925, Albert Sherbourne Le Souëf, the first director of Taronga Zoo in Sydney, identified as *A. forsteri* the skull and skin of a several seals that were collected at King Island, Tasmania, and Recherche Archipelago, Western Australia. However, for much of the early 1900s, the New Zealand fur seal was thought of as an occasional visitor to Australian waters. The breadth of its residence across South and Western Australia was not documented fully until Judith King's seminal publication on the identity of seals in Australia in 1969.

In the early 1920s, several researchers were formulating opinions on the taxonomy of otariids in southern Australia. Frederic W. Jones (who liked to combine his middle name with his surname, thus Wood Jones), the Professor of Anatomy at the University of Adelaide, realised that the most common fur seal in Australian waters had not been formally named. So in 1925, he described it, providing the name *Arctocephalus doriferus* (*dora* meaning skin or hide and *fero* meaning bear) for the Australian fur seal. Wood Jones' descriptions were flawed, though. First, as later surmised by Judith King, Wood Jones had based his description of an adult male on the skull of an immature female. Second, he based his behavioural attributes of the Australian fur seal on descriptions supplied to him by a ranger (C. J. May) from Kangaroo Island, where only New Zealand fur seals were likely to reside at the time.

Herbert Hedley Scott and Clive Lord from museums in Hobart and Launceston, respectively, rightly felt that the seal they were encountering in Tasmanian waters differed to the one described by Wood Jones. They named their seal *A. tasmanicus*, the Tasmanian fur seal. From then until 1969, researchers struggled to split the one species, and its range, into two. In 1969, Judith King recognised the errors in Wood Jones's account and clarified the situation. In recognition of priority, she used the species name *A. doriferus*. King also realised that the Australian fur seal was practically identical to the Cape fur seal, *A. pusillus* (Shreber 1776), of southern Africa. Ironically for the largest of fur seal, *pusillus* means small, and was applied only because the first description was based on a picture of a pup. In 1971, Charles Repenning (US Geological Survey) and others formerly recognised the Cape and Australian fur seals as subspecies, and named the Australian form *Arctocephalus pusillus doriferus*. They kindly attributed recognition of this species to Frederic Wood Jones, 1925.

The only change since the 1970s in nomenclature for the otariids of Australia has been the 2011 application of the genus name *Arctophoca* to New Zealand (and most other Southern Hemisphere) fur seals, by Annalisa Berta and Morgan Churchill. This recognised the marked differences between the Cape/Australian fur seals and all the other Southern Hemisphere fur seals.

There may be another change in the wind for the New Zealand fur seal too, but this is unresolved. In 2001, Louise Wynen (University of Tasmania) and colleagues performed the first detailed phylogenetic analysis of otariids. Amongst their findings was that the New Zealand fur seals' closest extant relatives were the South American (now *Arctophoca australis*) and Galapagos fur seals (*A. galapagoensis*). Sylvia Brunner from University of Sydney published in 2006 an extensive skull morphometrics study of otariids, which supported the findings of Wynen. Brunner proposed a subspecific status for these seals and offered '*australis forsteri*' for the New Zealand fur seal (*australis* coming before *forsteri* to denote *australis* was named first). There is some inconsistency in the phylogenetic relationships resolved by different studies, though. In 2007, Jeff Higdon (University of Manatoba) and colleagues preferred to retain the New Zealand and South American fur seals as separate species. Future studies are needed to resolve whether the New Zealand fur seal should remain *Arctophoca forsteri* or adopt *Arctophoca australis forsteri*.

As a final comment here, there has been some inconsistency in the literature with the use of the term 'seal'. This is a generic term that applies to all pinnipeds: phocids, otariids and odobenids (walruses). In the Northern Hemisphere, though, 'seal' sometimes is used to refer only to true seals (phocids), while occasionally in Australia (even in government documents), 'seal' can refer to fur seals. Thus, the phrase 'seals and sea lions' in Australia may refer to fur seals and sea lions (i.e. all otariids) while in the Northern Hemisphere it may refer to all pinnipeds except fur seals and walruses. To make it clear here, we refer to all pinnipeds as being 'seals'; and we use the term 'fur seal' for otariids that are currently in the genus *Arctocephalus*, *Arctophoca* and *Callorhinus*, and the term 'sea lion' for otariids in the genus *Neophoca*, *Eumetopias*, *Zalophus*, *Otaria* and *Phocarctos*.

3

MORPHOLOGY AND PHYSIOLOGY: ADAPTATIONS TO MARINE LIFE

Body shape

Seals need to travel quickly to catch prey and avoid predation, and a key feature for them to do this is a hydrodynamic body shape. Like many marine animals, seals have an overall spindle shape, being narrowest in diameter at the head and tail and broadest in the middle. Such a shape minimises resistance and drag through gradual separation then reforming of lamina water-flow over the body. External protuberances are minimised (e.g. ear pinnae) or retractable (e.g. female teats and the male penis and testes). Even limbs are partially internalised to enhance streamlining, although propulsion mechanisms – the digits of fore- and hindlimbs – are enlarged.

Compared with the average terrestrial mammal, seals are also large in size; this does not pose a gravitational problem in water. Large size helps to conserve heat and facilitates oxygen storage, which enhances diving capacity.

Pelage and skin

A cross-section of the outer integument of an otariid seal traverses long guard hairs; a fine underfur; a narrow epidermis; a dermal layer containing hair follicles, and sweat and sebaceous (secretory) glands; and a fat layer (blubber or hypodermis) prior to reaching underlying muscle (Figure 3.1).

In fur seals, the primary means of insulation is the dense guard hair and underfur. Guard hairs are oval shaped in cross-section and are regularly spaced. Each one is

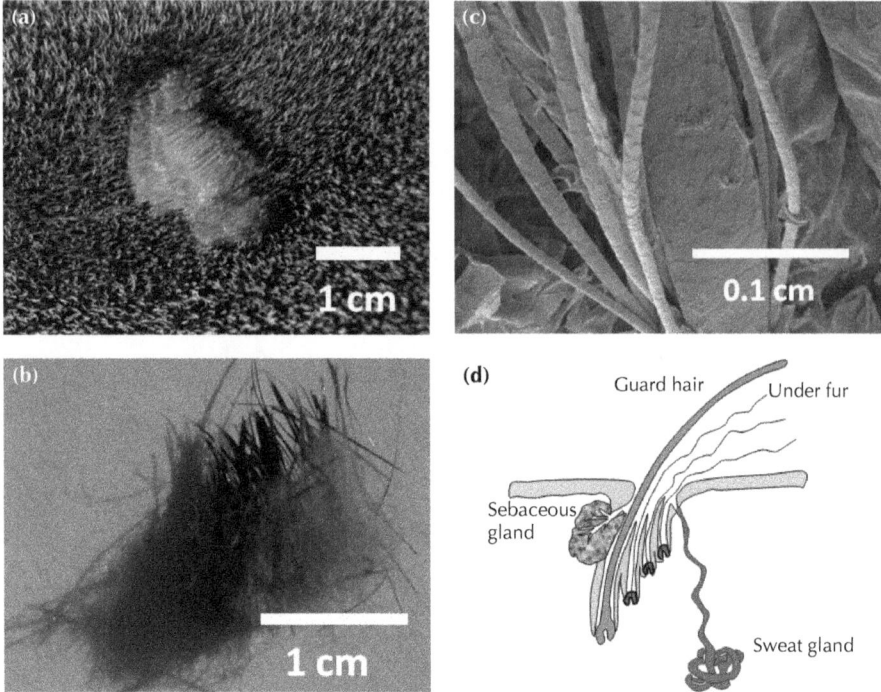

Figure 3.1. Pelage of the Australian fur seal: (a) external view with section trimmed; (b) trim of guard hairs (black with silver tip) and underfur (brown); (c) transmission electron microscope image of guard hairs and underfur; (d) cross-section of pinniped skin showing hair follicle and related structures. Photographers: (a, b) Roger Kirkwood, (c) Michael Lynch, (d) redrawn from King (1983) with permission; copyright Natural History Museum, London.

associated with up to 40 fur hairs. Combined, the hairs trap a layer of air against the skin while the seal is in the water and this provides great insulation. Sea lions (and phocids) lack a dense underfur. Their guard hairs may be sufficiently dense to prevent water from reaching the skin, but most insulation comes from a thick layer of subcutaneous fat. While ashore, seals spend considerable time scratching and grooming their outer integument.

All seals undergo a regular moult to renew their coats. For most, this is an annual process with a peak in hair turnover during late summer and autumn. Oddly, the Australian sea lion appears to have an 18-month moult cycle. The moult in fur seals and sea lions is gradual so they remain waterproof and do not need to fast over the moult. At the peak, though, individuals generally remain ashore for longer periods than usual in order to hasten the moult period. The surrounding terrain is often filled with wind-blown, loose hair. Adult females nursing pups are likely to forgo the comfort of the extended shore period, to maintain pup support. As a consequence, their moult is generally more protracted than the moult of males. In otariid seals, each hair is replaced individually. A typical moult commences around the face, then moves down the spine and around the body. In some phocids, such as elephant seals, moult takes place over a few weeks with the hair plus the outer skin shedding in clods. During such a moult, the seals prefer to remain out of the water.

Sebaceous glands are associated with each hair canal and secrete lipids onto the skin and hair. This conditions the hair, making it pliable and waterproof. Seals also have sweat glands which secrete moisture to aid in cooling when they are on land. Otariid seals have elongated nails (claws) on the middle three digits of their hindflippers; they use these nails extensively to groom the fur. The base of these nails can become clogged with fur during the moulting period.

Musculoskeletal structure and locomotion

At the early Miocene split between phocid and otariid seals, there was a divergence in locomotion strategies. Phocids took the evolutionary path of achieving in-water propulsion by sideways flexion of the backbone and hindflippers, whereas otariids developed foreflipper power. Consequently, basic musculoskeletal structure between phocids and otariids differs vastly (Figure 3.2).

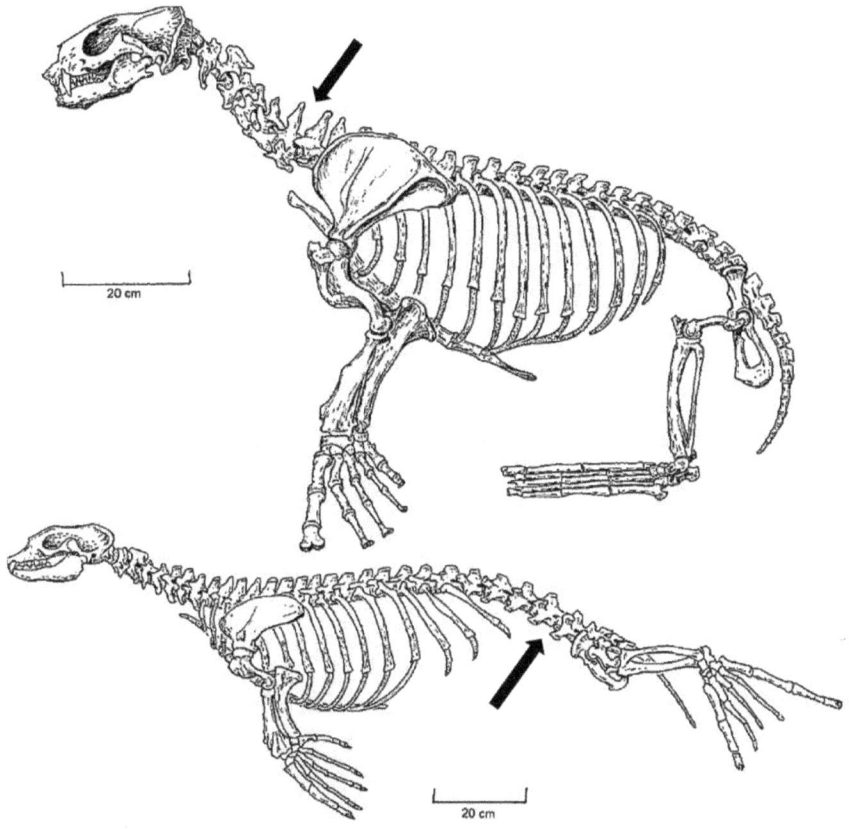

Figure 3.2. Lateral views of (a) an otariid skeleton (*Arctophoca forsteri*); (b) a phocid skeleton (*Monarchus tropicalis*). Note the prominent dorsal process (for muscle attachment) on cervical vertebrae of the otariid and the prominent transverse processes on the lumbar vertebrae of the phocid. These reflect the contrasting means of locomotion when in water (foreflipper versus hindflipper). Derived from King (1983) with permission; copyright Natural History Museum, London.

Synchronous movements of otariid forelimbs are levered by large brachial muscles (shoulder/chest to elbow). They do not possess a clavicle (collar bone), allowing a considerable arc of movement of the forelimb. The humerus (upper arm bone) is reduced in length and broadened compared with terrestrial mammals, providing greater leverage on the lower limb. It is also encapsulated in the body, reducing drag on the limb. The outer forelimb comprises webbed digits and a hardened integument (foreflipper). Stroke efficiency is enhanced by the foreflippers' hydrofoil shape, which provides hydrodynamic lift. The locomotory stroke is not a simple up-and-down motion, but incorporates backward thrusts and forward recovery phases, producing a figure-eight type movement of the limb.

The general shape of the otariid foreflipper varies slightly between species. Fur seals have adopted a narrow structure, which provides a high aspect ratio (length to width) and allows for great speed, whereas sea lions have a broader foreflipper which enables great manoeuvrability. Fur seals retain a high level of manoeuvrability through their generally smaller size, great backbone flexibility and rudder-like actions of the hindflippers. Digits of the hindflipper are webbed and may be splayed to provide drag for breaking and manoeuvring, or folded into a narrow, drag-reducing cone. The tail is a rudimentary stump.

Compared with phocids, otariids are particularly agile on land. This is because their flexible spine allows the hindflippers to tuck underneath them and the powerful forelimbs can hold their entire body mass off the substrate. The gait of otariids can involve individual or combined movements of both fore- and hindflippers. Thus, they may walk on all four limbs or gallop. With their low centre of gravity and area of contact with the substrate they also possess great climbing abilities and can access steep rocky shorelines. Phocids possess comparatively shorter forelimbs. They may either wriggle from side-to-side in a 'snake-like' movement utilising their forelimbs to grip and drag, or partially raise their body on the forelimbs, then throw body mass forward and progress in a 'grub-like' movement. The hindlimbs remain extended and assist little with locomotion on land.

Other features of the musculoskeletal system of seals are broad shoulder blades, an elongate sternum, broad ribs that protect lungs (which extend for much of the dorsal length of the body cavity) and a skull characterised by large eye orbits.

Sensory systems

Sensory systems enable animals to receive information from their surroundings. The acuity of each sense varies between species depending on its habitat, and ecological and social requirements. Senses of seals are attuned to their need to navigate and forage effectively in a marine environment, and to remain safe and socialise on land.

Touch – vibrissae

Touch receptors are distributed over the entire body of seals, but sensitivity to touch varies between species. Thigmotaxis, the propensity to touch other individuals, is stronger among sea lions than fur seals, although *A. pusillus* (Australian and Cape fur seal) is an exception. The thigmotactic behaviour of Australian fur seals contrasts with the desire for personal space of New Zealand fur seals (Plate 2a). Where these seals co-occur ashore, the tactile and larger bodied Australian fur seals take the smoother

Figure 3.3. Prominent facial vibrissae on a female New Zealand fur seal. Photographer: Roger Kirkwood.

substrate and the New Zealand fur seals take the sharper, uneven terrain where they will be left in relative peace.

Although most touch receptors of otariids are not merged, facial vibrissae (whiskers) are the exception. The vibrissae are stiff hairs associated with eyebrows (superciliary), the nose (rhinal) and the upper lip (mystacial). In pinnipeds, mystacial vibrissae are arranged in a series of rows, are heavily innervated (supplied with nerves), surrounded by blood sinuses (sacs) and controlled by voluntary muscles. They can sense sound compression waves, movement vibrations and water turbulence, and are likely to aid prey detection, particularly in low light conditions. In addition, they can detect changes in water currents or swim speed and assist navigation in low light, and on land can be used to judge close distances (e.g. when sniffing surfaces).

Vibrissae length varies across the pinnipeds. Walruses possess tough, stumpy ones that allow discrimination of the shape and size of potential food items, like bivalves, embedded in the substrate. Ringed seal vibrissae are intermediate in length; they contain 10 times the number of nerve fibres typically found in terrestrial mammal vibrissae and allow ringed seals to forage in very murky waters. Fur seal vibrissae are amongst the longest in the pinnipeds and New Zealand fur seal 'whiskers' can be particularly long (Figure 3.3). Long vibrissae could provide a large 'net' for detection of small, faster moving prey, such as krill and small fish.

Audition

The auditory (hearing) system is an important sense for seals both in air and in water. It enhances detection of the presence and direction of prey and predators, and enables communication between individuals. Seals detect sounds in a broad range, from 100 to 32 000 Hz.

External pinnae (the pinna is the outer part of the ear) represent a convenient way of deciding if a seal is an otariid or a phocid; otariids have them and phocids do not. The external pinnae of Australian fur seals stand out particularly prominently (Plate 2b). In otariids, the pinnae are capable of voluntary closure during diving; air trapped in the outer ear chamber is expelled under pressure.

Structurally, seal ears resemble those of terrestrial, generalist-hearing mammals. They possess a broad-bore external canal, a small ear-drum, a middle ear space encased in a tympanic bulla and an inner ear comprising a spiral cochlea with partial laminar (layered) support. While diving, tissues around the middle ear and external auditory meatus (cavity) fill with blood which replaces air spaces to equalise with external pressure. As the air space reduces with depth, hearing through the ear improves. Otariid ears differ from those of terrestrial mammals in having a large cochlea aqueduct, which amplifies sound reception. Their middle-ear bones are also more separated from the skull than are those of phocids. This reduces underwater sound amplification across the skull, but enables better discrimination of sound direction.

Phocid seals rely on vocal and auditory systems for underwater communication and produce sounds in a broad range, 100 to 15 000 Hz. Otariids also produce sounds when underwater, but have a smaller range, 1000 to 4000 Hz. They use clicks, grunts, whistles and releases of bubbles. On land, their vocalisations can resemble barks, growls, brays, trills and bleats. The sounds produced are species and age/sex-specific. Australian and New Zealand fur seals, for example, are readily distinguished by their unique calls.

Vocalisation is important for recognition of individuals across a busy colony. Otariids can identify differences in combinations of frequency, amplitude and timing of conspecifics from a considerable distance. Individual vocal detection is particularly important for maternal–offspring recognition. On return to the colony, an adult female produces her unique pup-attraction call. Several pups may identify the call as being similar to their mothers' and approach the female while emitting their unique mother-attraction calls. Females appear more discerning than the pups and ignore all calls but those most similar to that of their pup. Ultimate recognition of her pup relies on nose-to-nose smell and, upon sniffing at close range, recognition is instantaneous.

It was once thought that some otariids (and phocids) may be capable of echolocation, but this idea has been discounted.

Vision

Otariids rely on vision to sense their surroundings in two optically different media: air and water. On land, bright light levels are often encountered. In water, light is rapidly absorbed and scattered, and attenuates with depth; much of the foraging by seals is at very low light levels. Visual acuity is most advantageous in water and hence the eyes are better adapted to this medium. The eye positions of some species are more lateral-facing while others have more forward-facing eyes. In all, the eyes may be rotated to look directly upwards, which aids prey capture from below.

Seals possess a number of anatomical adaptations to the different pressures and light levels experienced in their semi-aquatic habit. Most obviously, their eyes are large relative to their body size and appear even larger underwater. They can resemble the eyes of nocturnal terrestrial mammals. The cornea has a highly keratinised epithelium for protection against wind, dirt and water. To limit light entry when in bright

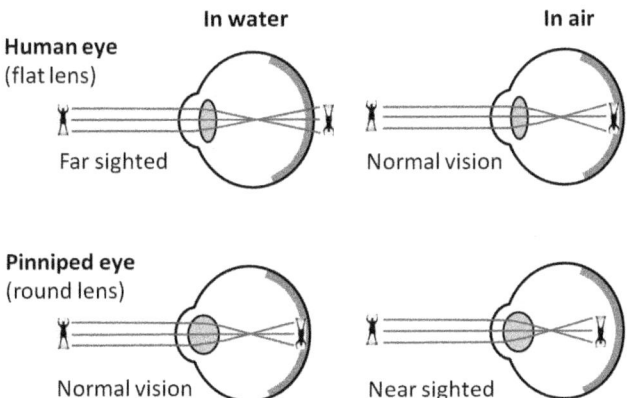

Figure 3.4. Comparison of the eye and lens shape of humans and seals, and optics in air and water. Redrawn from Berta and Sumich (1999) with permission, copyright Elsevier.

conditions, the iris can contract the pupil to a pinpoint. The powerful lens in the eye is near-circular in shape; this enhances underwater focusing, but the shape is difficult to adjust when out of the water (Figure 3.4). Powerful ciliary muscles (muscles comprising a ring of striated smooth muscle around the eye) can alter the shape of the lens, enabling focus at varying distances. Even so, on land seals are slightly myopic (near-sighted), which means distant objects are focused slightly in front of the retina and so appear blurred. At the back of the eye and behind the retina, a large cell layer (the tapetum lucidum) reflects light back through the retina and enhances low-light vision. A thickened sclera around the eyeball prevents deformity of the eye by pressure variations. Tear ducts irrigate the eye when the animal is out of the water, and tearing can be quite profuse if the eye is irritated by dryness, grit or bright surroundings.

Prey species of otariids are visually detected based on shape, movement and contrast – a dark shape against a light background. Otariid retinas are dominated by rods, which enhance sensitivity to dim light and nocturnal feeding, rather than cones, which allow colour discrimination. The few cones present are insensitive to wavelengths greater than approximately 500 nm. This means seals do not detect colour in the yellow to red end of the spectrum (Plate 3a). They do, however, possess dichromatic colour vision and can differentiate well between colours of blue–green wavelengths (400–500 nm). This represents an adaptation to underwater foraging where higher wavelengths of light are rapidly absorbed and only blue–green wavelengths are visible at any depth.

Although myopic on land, otariids clearly detect movement up to 100 m away, particularly if the movement is contrasted against a background (such as a skyline). Researchers find the least disruptive way to approach a seal is to wear substrate-coloured clothing, move slowly and carefully from downwind, and to squat or crawl.

Olfaction

Although otariids possess small olfactory lobes, there is substantial evidence that they have an acute ability to detect and differentiate odours. Nose-to-nose contact is an important means of communication and individual recognition. Adult female otariids

Figure 3.5. An Australian fur seal cow identifies her pup by its unique odour. Seals have a well-developed sense of smell. Photographer: Roger Kirkwood.

can identify their pup based on its unique odours (Figure 3.5). Olfaction is also important for adult males during the intense weeks of the breeding period. Fur seal bulls will detect and attempt to evict all other males, even juveniles, from their vicinity. They produce pungent exhalations which they can direct at neighbouring males. The odour communicates relative status to other males, aiding non-violent territorial defence. Females also take male odour as a cue when selecting whom to mate with. To detect odours that indicate a female is in oestrus, territorial males frequently sniff females themselves, as well as the air and the rocks where females have sat.

Another example of the strong odour detection of seals is that unfamiliar odours, such as boat exhausts, smoke or humans, can startle them, and cause some species of fur seal to flee to the water. Julia Back studied the responses of Australian fur seals to boat approaches and found that wind direction, carrying both sound and odour, strongly influenced how close a vessel could approach a site prior to its detection.

Taste
The tongues of otariids are short, wide at the back and taper to a curiously notched tip (Figure 3.6). They assist with swallowing, but are not utilised for panting or grooming.

Taste buds located on the tongue are chemical receptors that allow detection of dissolved substances that enter their mouths. Otariids have few tastebuds compared with terrestrial carnivores, suggesting a limited sense of taste. Experiments in captivity, though, have demonstrated that Californian sea lions do not detect 'sweet' tastes, but can detect 'sour' and 'bitter' tastes as well as humans can, and detect or distinguish 'salty' tastes 10 times better than humans. A high level of saline detection may be important for the otariid to recognise different currents and water bodies during navigation, and possibly for location of prey concentrations, by sensing the saline trails of excreted urea.

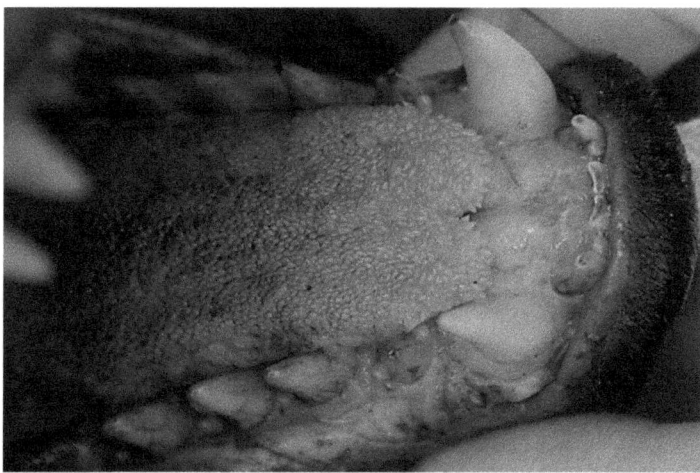

Figure 3.6. **The notched tongue of an Australian fur seal.** Photographer: Roger Kirkwood.

It is unclear, though, if taste is important in prey selection. Otariids swallow their food whole and in large chunks, suggesting taste alone is not an important prey selection factor. Seals display definite preferences for particular prey types, probably selecting them based on previous experience and texture.

Water balance (osmoregulation)

Seals can get all the water they need from their prey; fish and cephalopods consist of 60 to 80% water. However, individuals occasionally deliberately ingest small amounts of fresh or sea water. This can be important during periods of fasting, especially in warm climates, such as by otariid pups when their mothers are at sea, and by adult males during extended territorial tenures. If an excess of sea water is ingested, the stomach is upset and excess salts need to be eliminated, involving water loss.

Seals maintain their internal fluid levels below the salinity of sea water. Minimisation of water loss is crucial. Water is lost through exhalation, sweating and defecation, but primarily by urination. Relative to body size, the kidneys of seals (and other marine mammals) are larger than those of most terrestrial mammals. They comprise more than 100 discrete reniculi (lobes), each functioning as a miniature single-lobed kidney, with its own cortex, medulla, calyx and duct to the ureter (Figure 3.7). This multi-lobed structure is very efficient at extracting highly concentrated urea from the blood. In the animal kingdom, possession of a reniculate kidney is thought to be related to large body size and a requirement for highly concentrated urine.

To recover moisture from respired air, seals have an elaborate tubular bone system in their nasal passages. On land, they generally breathe through their nose, only using the mouth during exertion and when overheating. To minimise overheating, and concomitant water loss through sweating, panting and a higher metabolic rates, seals adopt a number of behavioural strategies. These include flipper waving, placing their hind flippers in a pool, throwing sand on their backs, resting in shade, moving to intertidal areas, and going for a swim.

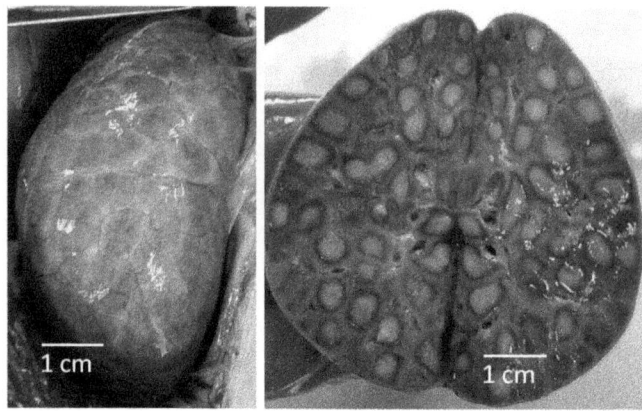

Figure 3.7. Example of a reniculate kidney, from a young Californian sea lion, *Zalophus californianus*: (a) *in situ*; (b) cut in half to demonstrate the shape of the reniculi. Permission to use photograph given by David Casper, UCSC Long Marine Laboratory.

Cardiopulmonary system

Pinniped cardiopulmonary systems are similar to those of terrestrial mammals. They breathe through their nose or mouth; the trachea is supported by cartilaginous rings; and their lungs are paired, lobate and terminate in tiny alveoli where gas exchange takes place. A feature of the pinniped respiratory system is that the diaphragm separating the thorax from the abdomen angles across the body and the lung extends for much of the dorsal length of the body cavity. An obvious difference near the heart compared with terrestrial mammals is an expansion of the aorta. This forms an 'aortic bulb', which acts to dampen blood pressure pulses and is believed to be an adaptation to diving.

Seals have exceptional diving capabilities. Elephant seals can breath-hold for up to 120 minutes and achieve depths greater than 2 km, and Weddell seals can breath-hold for 80 minutes and dive to 600 m. Although not amongst the diving elite, maximum diving breath-holds in otariids typically range between six and 16 minutes. Maximum dive depths of 180 to 500 m have been recorded. The vast majority of dives last under four minutes, though, and go no deeper than 100 m (Figure 3.8).

In human divers, long exposures to high levels of oxygen can be toxic and even short exposures to high nitrogen levels can be narcotic. Furthermore, pressurised gasses may dissolve in the blood and bubble out during ascent, causing blockages to circulation and tissue trauma (known as 'the bends'). Seals mostly avoid these issues by doing the equivalent of human 'snorkelling', taking air from the surface down rather than breathing in highly compressed air. Long-duration 'snorkelling' still has issues of pressurised gas absorption across alveoli walls. Seals overcome this by voiding air from their alveoli. First, they reduce air pressure by exhaling prior to descent. Second, alveoli collapse under pressure before bronchioles and the trachea, which are reinforced by muscle (phocids) or cartilage (otariids). Thus air remaining in the respiratory system is not trapped in the sensitive alveoli.

Seals do contend with the breath-hold limitations of hypoxia (tissues deprived of adequate oxygen), and accumulations of carbon dioxide and lactate in tissues, increasing the tissues' acidity. Attributes that enhance breath-hold ability over that of

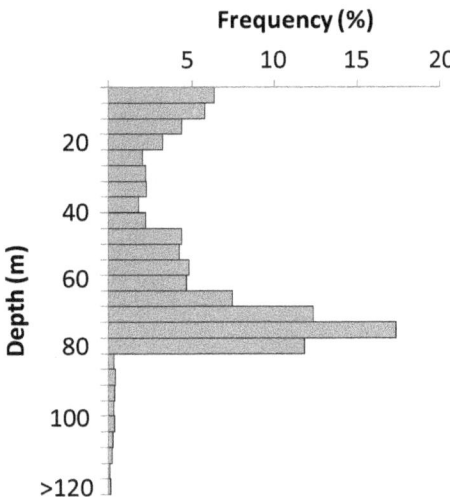

Figure 3.8. Frequency distribution of maximum dive depths (113 636) achieved by 13 female Australian fur seals from the Kanowna colony. Derived from Arnould and Hindell (2001). © 2008 Canadian Science Publishing or its licensors. Reproduced with permission.

terrestrial mammals include more red blood cells that contain more haemogloben (for oxygen carrying), greater storage of oxygen in blood and muscle rather than in lungs, a large spleen for storage of oxygenated blood, and a high acid-buffering capacity. These features contribute to extended aerobic (oxygen supported) dive limits. Aerobic dive limits can be predicted in seals based on their oxygen storage and metabolic rates. Seals can also push beyond their aerobic dive limits by employing tactics such as peripheral vasoconstriction (narrowing of vein diameters), which shunts blood preferentially to vital organs, and bradycardia (lowering of the heart rate). Interestingly, humans who train to dive for extended periods (up to six minutes) also undergo peripheral vasoconstriction and bradycardia.

Digestive system
The digestive system of seals is similar to that of terrestrial carnivores. Prey are grasped with pointed teeth and swallowed whole or torn into swallowable chunks by thrashing on the sea surface. Some seals also suck prey into their mouths, using a rapid withdrawal of the tongue, and then strain the prey against their teeth.

Phocid seals have 22–36 teeth, some possessing multi-cusped (multi-pointed) teeth, while otariids have 34–38 monodont (single-pointed) teeth. Like other carnivores, otariids have long canines and, in place of molars, have sharp, shearing, post-canines. All seals take live prey but some individuals may learn to scavenge (e.g. from trawl nets) or take dead prey while in captivity. The deciduous teeth of otariid pups are needle-like and are typically shed after 3–4 months. Erupted canines do not extend much beyond the length of the post-canines until late in the first year; this can be diagnostic for the age class. Otariids can have species-specific dentition so an inspection of teeth can aid species determination (Figure 3.9).

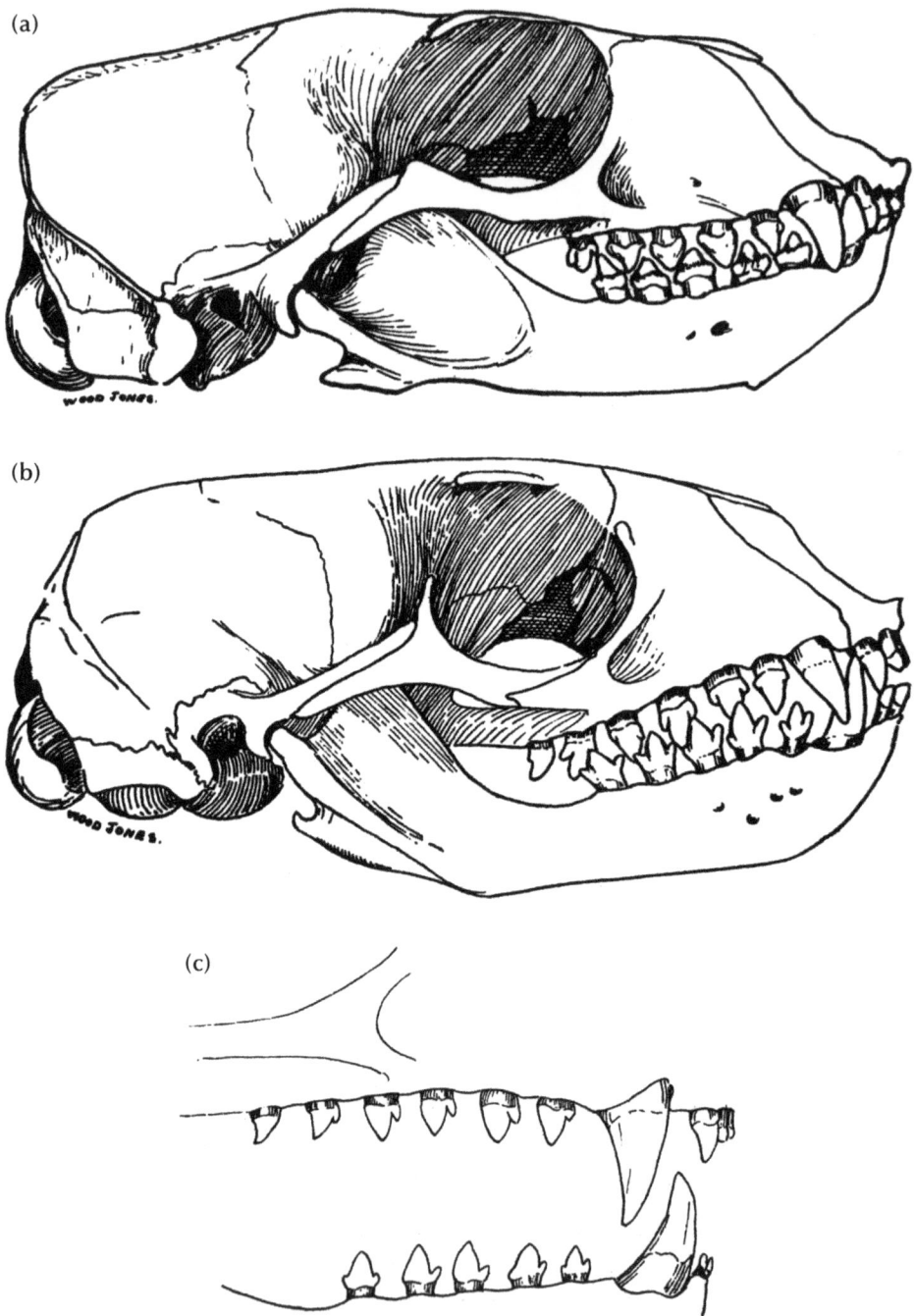

Figure 3.9. Comparison of dentition of the three otariids of southern Australia: (a) Australian sea lion; (b) Australian fur seal; (c) New Zealand fur seal. Reproduced from Jones (1925) with permission, copyright South Australian Museum.

Figure 3.10. Gastroliths collected from the stomachs of dead seals. The three on the left came from a single adult male Australian sea lion; those on the right came from a number of Australian fur seals. Photographer: Roger Kirkwood.

Salivary and oesophageal (mouth and throat) secretions aid the passage of food along a muscular and dilatable oesophagus to the seals' stomach. This is a single chamber. Food is chemically (acid) and mechanically (muscle action) broken down in the stomach, then passes through a narrow pyloric sphincter into the small intestine. A difference between phocids and otariids is that the pyloric end of the otariid stomach is sharply bent back on the body of the main stomach. The bend is termed an angular notch. This notch, along with the narrowness of the pyloric sphincter, serves to prevent passage into the small intestine of large particles, including larger fish bones, squid beaks and gastroliths (ingested stones), which must be regurgitated.

Many otariids swallow stones and retain these in their stomachs for a period. Australian sea lions are particularly adept at this and some stomachs examined have had up to 2 kg of gastroliths in them (Figure 3.10). The gastroliths may help the seals to sink (i.e. counter buoyancy), help stability, aid mechanical breakdown of food and/or reduce parasite loads. In adult males with extended territorial tenures, they may also reduce hunger pangs.

The small intestine is associated with a gall bladder, pancreas and large liver and is where most absorption takes place. The large intestine is short compared with the small intestine and only slightly wider in diameter. Seals have particularly long small intestines. Those of otariid seals are eight to 12 times their body length, compared with just four times the body length in terrestrial carnivores. Small intestines of some phocids can be 40 times the body length, similar to the ratio seen in terrestrial herbivores.

Digestion of food is rapid in otariids. Typical gut passage rates are 18–24 hours. Most digestion and defecation takes place at sea, so when on land the stomachs and digestive tracts are often empty.

Reproductive structures

Generally, reproductive structures of pinnipeds are comparable to those of terrestrial mammals. A difference, however, is the presence in the male of a baculum (penis bone or os penis), which is the ossified end to the corpus cavernosum. This may serve to protect the urethra during prolonged mating.

The reproductive cycle (discussed further in a separate chapter) is annual for most pinnipeds, but is about 17.5 months in the Australian sea lion. Adult female pinnipeds have three distinct phases of their reproductive cycle: oestrus, delayed implantation and foetal growth. Females come into oestrus post-partum; otariids remain in oestrus for 5–8 days while in phocids the timing of oestrus is linked to the duration of lactation (4–60 days). First-time breeders typically enter oestrus around the same time as adults of the species, although this can vary with occasional breeding in some species observed even several months prior to a normal breeding period. In adult males, testes are reduced in size for much of the year, then enlarge for the short period when females are in oestrus.

Externally, the sex of a pinniped may be determined from the position of the genital opening. This is posterior to the umbilicus in males and adjacent to the anus (so not visible externally) in females. The scrotum and testis of adult male otariids (but not phocids) may be evident outside the body cavity when on land, particularly during warm conditions. Female otariids possess four teats which are retracted when a pup is not suckling. Most phocids have two teats, whereas the walrus (like otariids) has four.

Sleep

When seals are ashore, they can sleep deeply. Some sleep so deeply that they do not notice when you walk right up to them. Generally, sleep is punctuated every 20 minutes or so by opening an eye (to check for danger) or for comfort behaviours, such as scratching and rolling over. Seals need time to recover from their exertions. By sleeping on land, they conserve energy, avoid predation and can socialise with other seals between sleeps.

Most seals can also sleep while at sea. Otariid seals commonly do this at the surface, holding one foreflipper in motion below, while the other is clasped between the hindflippers in a posture termed the 'jug-handle', which enables them to keep their nostrils clear of the water. The side of the brain controlling flipper in motion remains active, while the other side sleeps. The 'half-brain' sleeps may continue for hours, with periodic flipping (or could it be 'flippering'?) over to allow the other half of the brain to sleep (Plate 3b). Some seals may roll up in floating kelp, to aid flotation (and provide camouflage from predators) during sleeps at the surface.

In addition to surface sleeping, some seals may sleep during dives. Weddell seals can sleep during extended breath-holds while positioned just below the sea ice. Other seals may sleep while resting on the bottom or while drifting in mid-water. The ultimate sleep-drifter is the elephant seal, which may sleep for 20 minutes or more during 'drift dives' (also called 'recovery dives'). Elephant seals descend to a depth below which predators like killer whales normally venture, and then sleep-drift, sometimes performing a gradual spiral descent like a leaf falling from a tree. Martin Biuw from the Sea Mammal Research Institute, UK, and colleagues discovered that they could get an idea of an elephant seal's body condition by monitoring changes in the drift rate during 'drift dives'. Fat provides flotation and if the seal has more fat it is both healthier and sinks at a slower rate compared to seals in poor body condition. Seals in excellent condition may be positively buoyant, in which case they perform drift dives during the dive

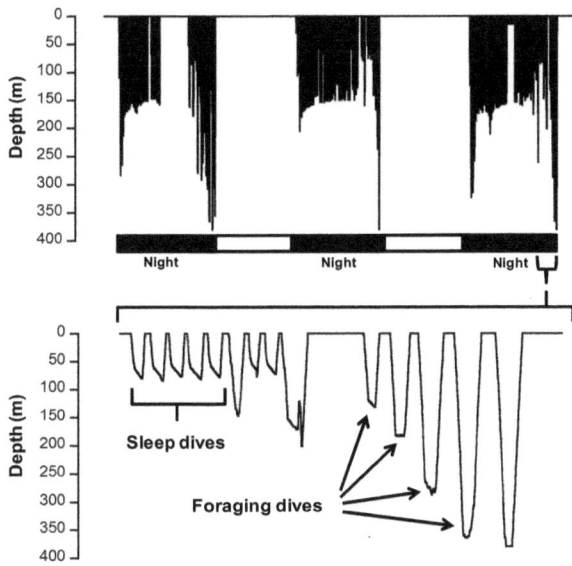

Figure 3.11. Profiles of dives performed by an adult male New Zealand fur seal. The upper figure demonstrates that most diving is nocturnal, with the deepest dives occurring at dusk and at dawn. The lower figure highlights a portion of inferred 'sleep' and 'foraging' dives. Derived from Page, McKenzie, Hindell and Goldsworthy (2005).

ascent. Monitoring the rates of depth change in drift dives across several months, the researchers could determine the locations where seals were doing well (putting on more fat, becoming more buoyant), and where they were doing poorly (putting on less fat, becoming less buoyant). Brad Page from the South Australian Research and Development Institute (SARDI) has also identified 3- or 4-minute drift dives in male New Zealand fur seals, suggesting that this behaviour may be more common among seal species (Figure 3.11).

In addition to sleeps, seals can take rests while at sea. For instance, most otariid species forage in bouts; that is to say, they perform a series of dives followed by a recovery period during which they rest near the surface. They particularly need such rests when they push themselves into anaerobic diving during foraging bouts, to recoup oxygen debt. Australian sea lions, however, appear to dive and forage continuously during foraging trips, which may last 3 or 4 days, rather than employ bouts and breaks. They always return to land to rest. This behaviour is thought to be an adaptation to reduce the chance of predation at sea, particularly by great white sharks (*Carcharodon carcharias*) which are common in the sea lions' foraging range.

This lack of resting at sea may explain why Australian sea lions can spend only about 50% of their time at sea, considerably less than the 70 to 80% of time being spent at sea by most other sea lions and fur seals.

Seals that forage in the open ocean may remain at sea for several months. Those that forage close to land (such as on continental shelf waters) tend to be at sea for shorter periods: either daily or on trips that average 4–7 days. When they return to land between trips, they spend much of their time sleeping.

Cognition

Seals are thought of as intelligent and playful animals that learn quickly from their experiences (Plate 4). A simple form of learning is habituation, which is the ability to modify a response through repeated exposure. Seals express habituation in their responses to benign stimuli. Initial exposure may seem a threat, but if harm does not follow, the stimulus may progressively be tolerated, then ignored. Seals may also express habituation in their foraging strategies, by returning to profitable areas and repeating techniques that improve prey capture and consumption rates. Where a food reward is obtainable, seals may even ignore potentially injurious stimuli. For example, during trials by David Pemberton and Peter Shaughnessy, Australian fur seals continued to visit fish farms despite the deployment of acoustic deterrents that had the potential to damage the seal's hearing, but otherwise caused them no harm.

Navigation

Fur seals and sea lions are likely to utilise a range of navigation techniques and cues for travel over short and long distances. For example, Catherine Wheatley and John Arnould reported at a conference in Hobart in 2012 that Australian fur seals could use man-made structures such as submarine cables for both navigation and foraging. Within their immediate vicinity, movement is directed by the sensors, vision, audition, touch (e.g. using vibrissae), and probably taste (chemical reception). Over greater distances, the most important cues to navigation are likely to be visual ones, such as movements of celestial bodies, dominant wind and current directions, and patterns on the sea floor (such as ripple marks in the sand or previously visited outcrops). Alternative means to aid direction of travel could be water temperature and salinity, and detection of sounds, such as boat engines or wave action on coastlines.

4

SEALS IN SOUTHERN AUSTRALIA

The three species of otariids that breed in coastal areas and islands of southern Australia (Australian fur seals, New Zealand fur seals and Australian sea lions, Figure 4.1) can occur in sympatry (overlap in ranges) and can be easily confused with one another. They possess common features and characteristics. For instance, they can all appear grey to brown with dark flippers when they are wet. Ultimate identification to species level can rely on attention to details and circulating photographs amongst experienced observers. The location and behaviour of the seal provide clues, as do size and head shape. Colour can also be useful, although it may be deceptive because an individual's colour changes when wet, throughout the year (depending on pelt condition and time

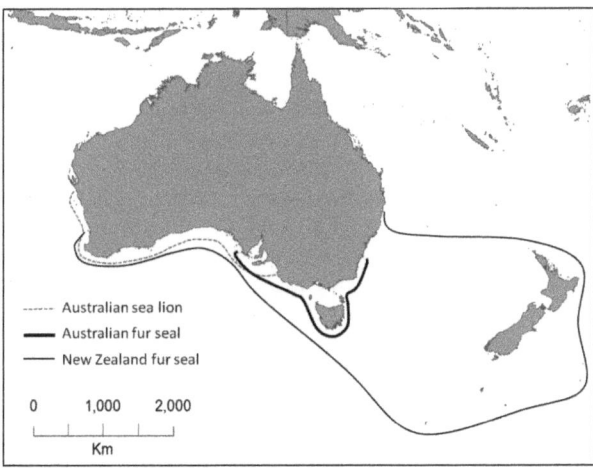

Figure 4.1. Ranges for the Australian sea lion, Australian fur seal and New Zealand fur seal.

Table 4.1. Seal species that may be sighted around continental Australia.

Family	Common name	Scientific name	Breeding area	Frequency in Australia
Otariidae	Australian sea lion[1]	*Neophoca cinerea*	Australia	Common
	Australian fur seal[2]	*Arctocephalus pusillus doriferus*	Australia	Common
	New Zealand fur seal	*Arctophoca forsteri*	Australia, NZ and subantarctic islands	Common
	Subantarctic fur seal	*Arctophoca tropicalis*	Subantarctic islands	Occasional
	Antarctic fur seal	*Arctophoca gazella*	Subantarctic islands	Rare
	New Zealand sea lion	*Phocarctos hookeri*	Subantarctic islands and NZ	Never
Phocidae	Southern elephant seal	*Mirounga leonina*	Subantarctic islands	Occasional
	Leopard seal	*Hydrurga leptonyx*	Antarctic sea-ice	Occasional
	Crabeater seal	*Lobodon carcinophagus*	Antarctic sea-ice	Rare
	Weddell seal	*Leptonychotes weddellii*	Antarctic sea-ice	Rare
	Ross seal	*Ommatophoca rossii*	Antarctic sea-ice	Rare

[1] Endemic species; [2] Endemic subspecies.

since moult) and with age. In the following sections, we explain characteristics of morphology, behaviour and location that help in the identification of seals.

As a first step to identifying a seal, the observer needs to determine what species could be present (Table 4.1), and if the one they are watching is an otariid or a phocid. Otariids have 'dog-like' faces, with a distinctly long snout and moderately long whiskers that regularly extend past the external ear pinnae. They move their foreflippers up and down to swim, and can walk and run on land, lifting their abdomen clear of the substrate. In comparison, phocids do not possess external ear pinnae, move in water using side-to-side movements of their body and travel on land by dragging their weight forward through propping on their foreflippers, or wriggling in a snake-like fashion.

If the animal observed appears to be an otariid seal, there are five or six likely candidates. The three most likely are the locally breeding species: the Australian fur seal, the New Zealand fur seal and the Australian sea lion (Table 4.2, Plate 5). Other possibilities include the subantarctic fur seal (*Arctophoca tropicalis*), which is a frequent visitor to the shores of southern Australia, the Antarctic fur seal (*A. gazella*), which is a rare visitor, and the New Zealand sea lion (*Phocarctos hookeri*), which has not been recorded on mainland Australia but individuals are regularly seen on subantarctic Macquarie Island.

Australian fur seal
Distribution
Australian fur seals are endemic to south-eastern Australian waters and breed at 15–20 colonies (Figure 4.2). Here, we define fur seal colonies as being locations where

Table 4.2. Comparative descriptions of the three otariids that breed in southern Australia.

	Australian sea lion	Australian fur seal	New Zealand fur seal
Growth			
Birth weight (kg)	6.4–7.9	5–12	4–6
Birth length (cm)	62–68	60–80	60–70
Weaning age (months)	15–18	10–12	8–12
Weight (kg): – male	180–250	218–360	120–180
– female	61–104	41–113	35–50
Length (cm): – male	185–250	201–227	150–250
– female	130–185	136–171	100–150
Longevity (years): – male	~25	~19	~15
– female	~25	~21	~22
Appearance			
Head	Large	Very large and broad	Large and broad
Brow	Little or no brow	Little or no brow	Distinct brow
Ear flaps	Small, against head	Long, stick out	Medium, against head, dark
Snout	Long, broad and blunt	Rounded (dog-like), bulbous nose	Pointy, bulbous nose
Whiskers	Medium	Long	Long, light colour
Pelage	Short hair, sparse underfur	Long hair, short thick underfur.	Long hair, short thick underfur
Colour: – males	Blackish-brown, cream-white wig	Light greyish-brown, pale chest, dark belly	Uniform grey to dark brown, pale muzzle
– females	Fawn to silver-grey, pale tan face and chest	Pale fawn to grey-brown, paler chest, brown belly	Grey to brown , lighter throat and chest
Pups	Dark brown, pale crown, dark facial mask	Black, grey ventrally	Dark brown, light snout and belly
Foreflippers	Paddle-shaped (rounded)	Paddle-shaped (rounded)	Oar-shaped (long with straight sides)
Hindflippers	Outer digits longer than middle digits	All five digits equal length	All five digits equal length
Reproduction			
Mature (years): – male	8–9	~5 (hold territories at 8–13)	4–5 (hold territories at 9–12)
– female	4–6	3–6	4–5
Implantation delay (months)	3.5–5	~3	~3

Table 4.2. Continued

	Australian sea lion	Australian fur seal	New Zealand fur seal
Gestation (months)	~14	~9	~9
Pupping interval (months)	17–18	12	12
Pupping season	Timing variable (5–7-month duration)	November–mid-December	late November–early January
Mating season	Timing variable (5–7-month duration)	November–December	mid-November–mid-January

15 or more pups have been recorded in one year. Other locations are referred to as either haul-out sites, or haul-outs with occasional pup births. Colonies mostly are located on islands that have easy access from the sea and have flat areas, gentle slopes, and/or boulder and cobble beaches. Almost 80% of pups are born at islands adjacent to the Victorian coast and 20% are born on Tasmanian islands of Bass Strait. There is a small mainland colony at Cape Bridgewater in western Victoria (40–50 pups in 2008–09) and one on North Casuarina Island, near the south-western tip of Kangaroo Island, South Australia (80 pups in 2011–12). Colonisation of North Casuarina represented a 50% expansion in the seal's latitudinal breeding range. Breeding distributions are likely to change in the future due to climatic and ecosystem change.

Australian fur seals rest at a further 50 or more haul-out sites across their range, including on man-made structures such as oil-rig struts and navigational structures. One major concentration of resting Australian fur seals is on islands in southern Tasmanian waters: Maatsuyker, Walker and Pedra Branca. These undoubtedly are productive feeding waters for this species. There are several haul-outs up the east coast of Australia as far north as Port Stevens, and in South Australia off Cape Jaffa, around Kangaroo Island and off the lower Eyre Peninsula. The species has not been recorded in Western Australia.

Australian fur seals forage almost exclusively over shelf waters of south-eastern Australia, although some off-the-shelf foraging has been recorded. The range extends from the Great Australian Bight in the west, around Tasmania and to northern New South Wales. When on land, Australian fur seals are particularly skittish and will stampede into the water if disturbed by an unfamiliar sight, sound or smell. Offshore rocky islands and isolated peninsulas afford the most secure resting locations. The seals also access large coastal caves (such as at Cape Bridgewater, Cape Nelson and Cape Volney in western Victoria). Individuals occasionally come ashore alone to rest where they think it is safe to do so, but tend to avoid sand beaches unless they are unwell.

Description

Australian and Cape fur seals (*Arctocephalus pusillus* spp.) are eminently distinguishable from other Southern Hemisphere fur seals (*Arctophoca* spp.) by their sea-lion-like characteristics. They vocalise like a sea lion, rest in contact with conspecifics as do sea lions (thigmotaxis), and are almost twice the size of the other fur seals. Females and

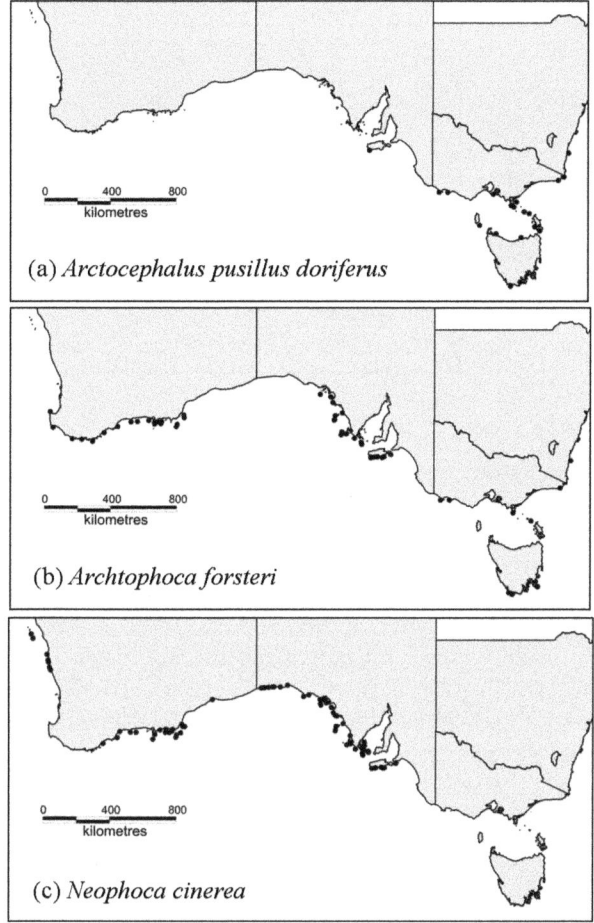

Figure 4.2. Breeding sites for (a) Australian fur seal; (b) New Zealand fur seal; (c) Australian sea lion.

males weigh up to 120 and 350 kg, respectively, compared with maximum weights of 50 and 150 kg for the other Southern Hemisphere fur seals (Figure 4.3).

Like other otariids, Australian fur seal pups vary in colour depending on their age (Plate 6). They are born with black guard hairs, a variable grey underside and a coat of fine, light cream underfur. The pelage develops on the neonate and grows quickly from birth with guard hairs reaching 1–2 cm in length. Pup guard hairs are softer than hairs of older individuals and clump when they emerge from the water. Over several weeks, their hair fades to a dark grey with the belly and chest hairs taking on a fawn colouration. Live births occur at colonies and haul-out sites from around August, but those born before November typically succumb within hours. Most births occur between mid-November and mid-December. Black pups will readily enter the water within days of birth, but endeavour to remain in pools or less than a few metres from the shore.

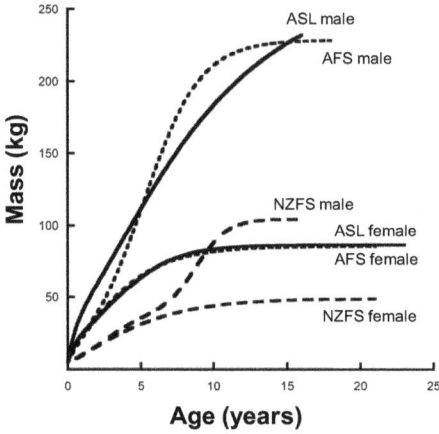

Figure 4.3. Age and mean body mass of Australian sea lions, Australian fur seals and New Zealand fur seals. The age at the end-point of the curve is the estimated longevity. Derived for the respective species from McIntosh (2007a), Arnould and Warneke (2002) and McKenzie *et al.* (2007).

During summer, there is a high mortality of pups, due to starvation (inexperienced mothers), injury and drowning, and potentially through diseases. This is a natural and annual occurrence, but the number of pups succumbing each year varies enormously, depending on weather and location-dependent factors. Dead and dying individuals often come ashore on public beaches near colonies.

In February, Australian fur seal pups moult into a silver-grey pelage. Colour-wise, they are then indistinguishable from juveniles and are readily confused with them. A common mistake is to think an independent juvenile is a pup. Pups tend to gallivant around with enthusiasm and a lack of coordination while juveniles are more muscular and 'serious'. Compared to juveniles, pups can have a 'chubby' appearance, because of being nursed rather than having to hunt for fish. Some cows may suckle young into their second and third year, however, and juveniles that continue to receive milk do retain a pup-like chubbiness. Definitive distinction between pups and juveniles is evident in their teeth. The lower canines of pups are about the same size as their incisors and post-canines whereas those of juveniles are at least twice the length of their other teeth. Seals frequently open their mouths so with patience an observer can make the distinction without having to capture an animal.

Weaning is a poorly understood process that occurs between September and October for most individuals. It is a tough time for some pups to switch to independence and many die. Each year, many weak individuals land on public beaches and linger around jetties seeking easy meals, and draw concern from members of the public. The 'yearlings' have distinctly short canines compared with the canines of older juveniles, until about the following February.

Juveniles are 1–3-year-old seals of either sex. They are much the same colour as the adult females – silver-grey to brown, with lighter hair on the face, throat and chest, and with a light-reddish-brown underfur. Individuals in poorer condition can have an overall brownish appearance, which may indicate a disruption or delay to a normal moulting regimen. Unlike pups, juveniles may travel hundreds of kilometres from their natal sites, and rest at other colonies and haul-outs, so may be encountered across the species range.

Adult females are a light greyish-brown with a pale-grey chest, brown belly, and reddish-brown underfur. When wet, they will appear uniform grey, with the paler chest. The head is smaller and narrower than that of the male, with no obvious brow, and they have a narrower neck and chest. Subadult and adult male Australian fur seals are similar in colour to the adult females but a little darker. Adult males have a 'mane' of up to 5 cm long hair on the nape, chest and shoulders. The nape may be lighter in colour in older animals and resemble the 'wig' of adult male Australian sea lions.

New Zealand fur seal
Distribution
The breeding distribution of the New Zealand fur seal in Australia is centred in South Australia, where about 80% of pup production occurs at five main colonies: North and South Neptune islands, Liguanea Island, and at Cape du Couedic and Cape Gantheaume on the south coast of Kangaroo Island. There are numerous minor colonies scattered on other offshore islands off the western Eyre Peninsula. There are also breeding colonies in Western Australia (centred on the Recherche Archipelago), and around Victoria and Tasmania. In New Zealand, this species breeds on rocky headlands of the South Island and off the south-east corner of the North Island, on Stewart Island and on subantarctic islands: Snares, Campbell, Chatham, Auckland, Bounty and the Antipodes.

New Zealand fur seals also frequent many haul-out sites throughout their range. This includes subantarctic Macquarie Island (south of Tasmania) and up the coast of New South Wales. Pups may be born on occasions at these haul-outs. New Zealand fur seals forage pelagically (near the surface or in the water column) and can cross vast distances of open ocean. For example, individuals marked in New Zealand have come ashore in South Australia, and vagrants have turned up on tropical shores, such as in New Caledonia.

Description
New Zealand fur seal pups are born dark brown to black with a lighter snout and belly. Pups appear 'fluffy' when dry due to their very dense underfur propping out the long, silver-tipped guard-hairs. Compared with Australian fur seals, New Zealand fur seal pups are more petite, with smaller, pointy foreflippers. The pups moult their pup coat (lanugo) when about 4 months old and grow a silver-grey pelage similar to juveniles. They wean at about 8–10 months of age.

All animals are a metallic grey immediately after moulting (March to May), then gradually become browner through to the next moulting period. The exception to this are 'yearlings' (juveniles 12–24 months of age), which are rarely sighted in colonies and haul-outs. Most may not fully moult in their second year, giving their coat a rusty brown appearance. Juveniles are a rich dark brown with a pale white or cream 'moustache'.

In adult females, the chest and throat are light brown, and the abdomen is dark brown. The head is smaller and narrower than that of the male, with no obvious brow. Adult males are uniform dark grey to brown with pale muzzle fur. Like Australian fur seal males, New Zealand fur seal males may be distinguished from females by their larger head, and heavier neck and chest.

It is difficult to distinguish between Australian and New Zealand fur seals based on colour as both can appear to be silvery/grey/brown. However, the Australian fur seal can be browner, lighter coloured and more mottled, compared with the New Zealand fur seal, which may appear more uniform dark grey to dark brown in colour. The New Zealander's coat appears 'thicker', its head and snout are narrower and more pointed, and its vibrissae are white and can get very long (sometimes exceeding head length). Also, its eyes face distinctly forward compared to a more lateral eye position for the Australian fur seals. Post-canine teeth provide a key diagnostic distinction between the species, with those of Australian fur seals possessing anterior and posterior secondary cusps; post-canines of New Zealand fur seals lack posterior secondary cusps. The teeth of New Zealand fur seals also appear to be more spaced than the teeth of Australian fur seals.

Australian sea lion

Distribution

Australian sea lions are Australia's only endemic seal species (the Australian fur seal is an endemic subspecies). The breeding distribution of the Australian sea lion extends from the Houtman Abrolhos Islands off the west coast of Western Australia to The Pages Islands, just east of Kangaroo Island in South Australia. Around 85% of the pups are born in South Australia. No pups are born in the eastern Australian states although older individuals occasionally roam into Bass Strait and up the east coast of Australia. Prior to sealing, Australian sea lions were common in Bass Strait, although it has not been confirmed if they pupped there.

Due to its unique breeding strategy, an Australian sea lion site is referred to as a colony if more than five pups are recorded (contrasting with the 15 pups required to recognise a fur seal colony). Pupping has been recorded at 78 sites: 50 in South Australia and 28 in Western Australia. Despite the large number of breeding sites, only eight produce more than 100 pups per breeding season, and all are in South Australia (North Pages, South Pages, Seal Bay on Kangaroo Island, Dangerous Reef, Lewis Island, West Waldegrave Island, Olive Island and Purdie Island). The average pup production per breeding site is just 48. The largest colony, Dangerous Reef, is more than twice the size of any other breeding colony, and produces more pups than all of the Western Australian colonies put together.

Another 151 locations have been identified as haul-out sites: 90 in South Australia and 61 in Western Australia. Because records of haul-out sites are based on opportunistic observations, the actual number could be higher than this.

Description

Australian sea lion pups are dark grey at birth. Their coat becomes brown with pale belly fur by about 4 weeks. Pups have usually fully moulted to an adult female-like pelage by about 5 months. Juveniles and adult females have marked countershading, darker dorsal fur and pale abdominal fur, which could be an adaptation that reduces shark predation. Their dorsal fur is silvery-grey after moult and gradually fades to brown. The pale creamy white fur on the chest and abdomen extends to the sides of the face and above the ears.

Juvenile and subadult males often have dark spots on their chest; this distinguishes them from females. As males age, these spots become larger and more numerous until the chest is completely dark. Adult males are grey (post-moult) to dark brown and black all over, with the exception of a creamy white cap extending from the eyebrow to the lower neck and shoulders. Compared to females, the male neck and shoulders are greatly enlarged. The large head has a long, broad snout that is blunt in profile. Pale whiskers extend just past the small ear pinnae. The pinnae lie close to the head and contrast with fur seals' pinnae, which tend to stick out.

Visitors

Australian continental waters also are within the foraging ranges of seal species that breed elsewhere (Plate 7). Individuals of these species that haul-out on Australian shorelines often (though not always) are in poor health. This may be because they are foraging in less optimal habitats near the edge of their range, or the poor health may have impaired their navigation skills, influencing their arrival at shores they would not normally visit.

The phocids that most commonly visit continental Australian shores are southern elephant seals, which are large and brown with big eyes, and leopard seals that are mottled black-grey and possess '*Tyrannosaurus*-like' jaws. Other phocids that rarely come ashore are crabeater, Weddell and Ross seals. Amongst the otariids, subantarctic fur seals are the most common visitors to Australian shores, whereas Antarctic fur seals and New Zealand sea lions are rarely, if ever, seen.

Southern elephant seals

Southern elephant seals breed on sand and shingle beaches of subantarctic islands (including Macquarie and Heard), islands adjacent to the Antarctic Peninsula and at Peninsula Valdez in Argentine Patagonia. They range throughout the Southern Ocean and northwards to the coastlines of all Southern Hemisphere continents. Colonies were present on the north-west coast of Tasmania (eliminated perhaps by Aboriginal hunters 2000 years ago) and on King Island, Bass Strait (eliminated by sealers in the early 1800s). Elephant seals routinely haul-out to moult at sites across their range, including on the coast of Antarctica.

Elephant seals are readily distinguishable from other phocid visitors to southern Australia by their blunt snout. When in breeding condition, adult males possess a large, fleshy 'trunk'. Other distinguishing features include large eyes, short, sparse teeth that appear to just protrude through the gum, brown coat and large size.

Elephant seals have an annual reproductive cycle. Breeding-sized adult males (over 10 years old, 3.5–4.5 m in length and 2000–3800 kg) gather at breeding beaches in September prior to the arrival of cows (over 3 years, between 2 and 3 m, 250–800 kg). The biggest males, termed 'beachmasters' (typically over 14 years old), hold harems that may exceed 100 cows. In September–October, pregnant cows come ashore and give birth to a single, 45 kg pup, which is suckled for approximately 23 days. Prior to departure, the cow mates with the beachmaster (and maybe opportunistic challengers). Weaned pups weigh about 110 kg on average (up to 250 kg) and remain ashore fasting for a further 5 weeks. During that time, they moult and convert stored fat into lean tissue. Through spring, summer and autumn, there is a constant progression of moulters: first juveniles, then subadult males, adult females, and lastly adult males. It takes each seal 30–40 days to complete its moult. They avoid entering the water and do not feed during this period. Longevity for an elephant seal is 20–25 years.

Mark Hindell from the University of Tasmania and Harry Burton from the Australian Antarctic Division have led research into southern elephant seals at Macquarie Island and Heard Island since the 1980s. Amongst the many fascinating findings has been the recording of the deepest and longest duration dives for any pinniped, 2 km deep and over two hours, respectively. Most dives, however, are to depths less than 500 m and for less than 25–30 minutes. Elephant seals spend 90% of their time at sea submerged, only returning to the surface for fleeting, two minute reoxygenations. Southern elephant seals eat mainly myctophid (lantern) fish and squid, which they target at depths that are usually more than 200 m.

During the 1800s, elephant seal populations on Macquarie Island and Heard Island were harvested to near extinction. By the 1960s, the populations had recovered to approximately 140 000 on Macquarie and 50 000 on Heard, but between 1960 and 1990 both suffered declines of approximately 50%. Curiously, the declines did not affect elephant seal populations of the south Atlantic. Declines in the southern Indian and Pacific Oceans possibly related to changes in distribution and abundance of the seal's prey. Since the 1990s, the populations have remained relatively stable.

In southern Australia, elephant seals are most likely to come ashore between September and March to rest and moult. Those coming ashore to moult are liable to remain for up to 30–40 days. Through this time, they occasionally enter the water and shift between beaches. Several elephant seal pups have been born on beaches around Australia, notably in Tasmania, and usually in September.

Leopard seals

Leopard seals breed on the Antarctic pack ice in spring. Like southern elephant seals, they range throughout the Southern Ocean and northwards to all continents of the Southern Hemisphere. They are the phocid seal which is most commonly seen on southern Australian shorelines.

Leopard seals are easily distinguishable from other phocids by their large, reptile-like heads and humped shoulders. They have a huge gape that displays substantial, tri-cusped teeth. The body shape of leopard seals is more slender than that of other phocids in their range and their foreflippers are longer. Each leopard seal has a unique pelage, which is a mottled, spotted grey colour that typically is darker on the dorsal surface, but all have a broad pale band above their upper lip.

Female leopard seals are slightly larger than males (up to 600 kg compared with up to 400 kg). They give birth between October and December to single pups which they nurse for about 30 days. Both sexes 'sing' to attract mates and mating events occur in the water soon after the pup is weaned. Tracey Rogers from Taronga Zoo and the University of New South Wales researched the singing behaviour of leopard seals. She found that females may sing for 3 days during the breeding season, whereas males sing almost continuously through the breeding period to maintain territories and maximise mating opportunities.

The diet of leopard seals varies depending on what is available to them. Antarctic krill can form the bulk of the diet in some areas and fish are important in others. Some leopard seals will specialise at ambushing penguins around colonies, especially fledging chicks. Recently weaned seals can also be taken. On rare occasions, leopard seals have bitten humans. On one occasion, a researcher snorkelling in waters near Rothera Station on the Antarctic Peninsula was taken under water and drowned by a leopard seal. The leopard seal possibly assumed humans were odd seals or penguins.

Sightings of leopard seals on southern Australian beaches occur year round, but are most frequent in late winter and spring. The number of leopard seals visiting fluctuates between years and erupts periodically, possibly in relation to episodic northward extensions of Antarctic water. Individuals may remain ashore for several hours to several days. Some may traverse around the coast for several months, feeding on fish, squid, little penguins and shearwaters, and hauling out occasionally to rest.

Crabeater seals

Crabeater seals breed on the Antarctic pack ice. With an estimated population of 12 to 50 million, they are by far the most abundant pinniped in the Southern Hemisphere. Although largely associated with sea ice, they range across the Southern Ocean and occasionally into temperate, continental waters. There are several records of individuals hauling out on beaches in south-eastern Australia.

Crabeater seals can be confused with leopard or Weddell seals, but are distinguished by their dog-like snout and cream-grey to light brown pelage (this may be mottled near the flippers). Also distinguishing them is that crabeaters often have moderate scaring around the head and chest from fights with conspecifics and deep scaring along the body resulting from close encounters with leopard seals. A further feature is that crabeaters on land regularly wriggle in a snake-like fashion, using their foreflippers to grip alternately, whereas other southern phocids typically hump along using their foreflippers.

Female crabeaters haul-out on sea ice to pup in early October. Often, they are accompanied on the ice by one (sometimes more) males that wait for the female to come into oestrus. This occurs when the pup weans at approximately 20 days after birth. Outside of the breeding period, crabeater seals are more social than are leopard seals. Groups of young males in particular will haul-out to rest on the same icefloe.

Really, crabeaters should have been named 'krilleaters'. They feed almost exclusively on Antarctic krill and are unlikely to ever see crabs. Their name was given to them because the krill that poured out of their stomachs when sealers opened them resembled crab meat. Most foraging for krill occurs off the Antarctic continental shelf and in the top 50 m of the water column, although crabeaters can dive to 500 m depth.

Weddell seals

Weddell seals are the most southerly living pinniped species and their lives are strongly linked to the sea ice. With such strong links to the sea ice, it is surprising that some also venture into temperate waters. Individuals have been recorded as far north as Uruguay and New Zealand, and one was caught in fishing net in South Australia in 1913.

Like leopard seals, Weddell seals have a mottled, individually unique, pelage. They can be distinguished from leopard seals by their smaller head in relation to body size and from crabeater seals by their shorter snout and full-body mottling. Also, Weddell teeth are single-cusped; leopards and crabeaters have tri-cusped teeth.

Early in October each year, pregnant female Weddell seals haul-out onto the ice, usually the fast ice which they access via cracks caused by tidal movement. The scarcity of suitable tide cracks in October probably contributes to Weddell seals being more colonial breeders than the other sea ice seals. Pups wean at between 6 and 7 weeks. During the later weeks of suckling, pups can enter the water and perform shallow dives. They start to supplement their milk diet with prey. Adult males establish underwater territories around the tide cracks and mate with females late in lactation. Weddell seals moult in summer, and can gather in large numbers (100 or more) at suitable moulting sites, such as permanent snowbanks on shorelines.

Rob Harcourt from Macquarie University investigated the diving behaviour of Weddell seals, which are second only to elephant seals in diving capability. They may descend to 600 m and remain submerged for over 70 minutes, although dives usually are to less than 200 m and for under 10 minutes.

Since 1973, diet and demographic studies of Weddell seals at the Vestfold Hills, Antarctica, have been led by Harry Burton, Samantha Lake and others from the Australian Antarctic Division. The research has also provided details of the seals' prey, movement and breeding biology. Weddell seals in the Vestfold Hills prey mostly on fish (such as *Pleuragramma antarcticum*), but also on cephalopods and crustaceans (particularly prawns). The seals tend to remain within the fast-ice zone year-round and have a high degree of interyear fidelity to breeding areas.

Ross seals

Ross seals are the smallest, least abundant and least studied of the Antarctic phocids. They are solitary and usually sighted in sea-ice regions of the Southern Ocean but several have been sighted on subantarctic islands and, in September 1953, a juvenile female came ashore at Beachport in South Australia.

Diagnostic features of the Ross seal are its relatively thick throat and its pelage, dark brown on the dorsal surface and fawn on the ventral surface. Its unique response to approaching danger is to raise the head, bulge the throat, open the mouth wide and emit a series of unusual trilling, clicking, cooing and thumping sounds.

Colin Southwell and colleagues from the Australian Antarctic Division conducted research on Ross seal diving behaviour and population status as part of an extensive survey during the 1990s and early 2000s of sea-ice seals in the Australian Antarctic Sector. They found that one reason the seal is infrequently seen is that it spends proportionally more time under water than any other pack-ice seal. Females give birth on the pack ice in late October–November. The pup may enter the water within days of birth and wean at 3 or 4 weeks of age. Ross seals regularly dive to depths of 150 m. Their main prey appears to be squid, although fish and krill are also eaten.

Subantarctic fur seals

The major population centres for subantarctic fur seals are Gough Island in the south Atlantic, and Prince Edward, Marion and Amsterdam islands in the southeast Indian Ocean. In the mid to late 20th century, the species colonised (or recolonised) Iles Crozet and Macquarie Island. At Macquarie Island, subantarctic fur seals interbreed with Antarctic and New Zealand fur seals. They are also the most common 'visiting' otariid of the southern coastline of Australia. Most individuals that come ashore around southern Australia are juveniles although adults also occur.

Adult males are perhaps the most easily recognisable, possessing a chocolate brown to black dorsal fur with contrasting creamy-yellow chest and face, with a crest of black fur on top of head that becomes erect when excited. Females have a similar pelage to males, without the crest. The colouration can be fainter in juveniles, making them easily confused with other species, especially the Antarctic and New Zealand fur seal. Often it takes a circulation of photos around experienced observers to correctly identify these. In general, subantarctic fur seals have a shorter, snubbier snout (akin to a Pekinese dog), and more compacted and shorter flippers compared to other fur seals.

The breeding biology of subantarctic fur seals is comparable to that of other fur seals. Bulls guard territories that females pup within (during December and early January), and females then suckle pups between foraging bouts at sea until pups are weaned (when about 10 months old). Subantarctic fur seals mostly feed on myctophid fishes.

Macquarie Island was discovered by sealers in 1810, and they had exterminated the fur seal population within a decade, harvesting over 200 000 skins. Sealers left little evidence of which fur seal species were present. Post-sealing, no fur seals bred on the island for at least 130 years, until 1954.

Peter Shaughnessy (CSIRO, then South Australian Museum) and Simon Goldsworthy have monitored the recovery of fur seals at Macquarie Island since the mid-1980s. The recolonisation of Macquarie Island by fur seals has been slow and complex, with three species involved: the New Zealand, Antarctic and subantarctic fur seal. The islands' great distance from major population centres of seals has meant that immigration rates have been low, and colonisation by males and females from each species has been asynchronous. This has led to extensive hybridisation, the highest reported among pinnipeds. Antarctic fur seal females colonised first, in the mid-1950s, but males of this species did not hold breeding territories until the late 1980s. Antarctic fur seal females were most likely mated by New Zealand and subantarctic fur seal males, which had been visiting the island since at least the 1940s and 1960s, respectively. Subantarctic fur seal females commenced pupping on the island in the early 1980s, while New Zealand fur seal females have not yet pupped there (data up to 2010). Although pups were recorded in 1954, few births occurred in most years up to around 1980. Since then, births have increased at about 6% per year and 235 pups were born in 2010. Genetic studies by Melanie Lancaster (PhD, La Trobe University) in the early 2000s indicated that about 52% of pups born at Macquarie Island are Antarctic fur seals, 22% subantarctic fur seals and 16% hybrids. The proportion of hybrids had decreased from 30% in the early 1990s, mainly due to the availability of conspecific mates for Antarctic fur seal females. Pre-mating isolating mechanisms, such as female mate choice and habitat preference, also helped to reduce hybridisation levels.

Antarctic fur seals

Antarctic fur seals are the most abundant of the otariids in the Southern Hemisphere with a population estimated to be over six million individuals during the early 2000s. They have a circumpolar breeding distribution on subantarctic islands and islands off the Antarctic Peninsula, but with 80% of the species being based at South Georgia in the south-west Atlantic Ocean. Small populations colonised Heard and Macquarie islands during the late half of the 1900s. Individuals have not been recorded around continental Australia.

Antarctic fur seal pups are born mostly in early December. Unlike the 10-month suckling period of most other fur seals, Antarctic fur seal pups are weaned after 4 months. After the breeding season, females depart and mostly remain at sea until they return to pup the following summer. Non-breeding males tend to come ashore to moult at the end of the breeding season.

At Heard Island, following elimination by sealers in the mid-1800s, fur seals started to return in the 1950s. Two pups were sighted in 1963, increased to 248 by 1987 and to approximately 1500 in 2003. In addition to the breeding population, large numbers of male Antarctic fur seals, over 30 000 in 2001 for example, come ashore in late summer to moult. These males probably originate from the much larger breeding colonies at Iles Kerguelen, north of Heard Island. Antarctic fur seals eat krill in south Atlantic waters of the Southern Ocean, but eat mostly myctophid fish elsewhere.

New Zealand sea lions

New Zealand sea lions (*Phocarctos hookeri*) have a small population that is centred at the Auckland Islands, south of New Zealand. They also reside at other New Zealand subantarctic islands and on the coast of the South Island of New Zealand. Juvenile and subadult males are regular visitors to Macquarie Island. Individuals have not been recorded around continental Australia.

5

REPRODUCTIVE BIOLOGY

Breeding sites

Otariid seals haul-out on land for many reasons and generally are gregarious when ashore. A prime requirement of haul-out sites is that other seals are present. Otariids in southern Australia generally prefer rocky shores, although Australian sea lions do utilise sand beaches, notably at Seal Bay on Kangaroo Island, South Australia. Cape fur seals of southern Africa also haul-out in large numbers on sand beaches, which suggest that the conspecific Australian fur seal is capable of doing this, although it does not at present. In Australia, fur seals resting on sand beaches are often unwell. Individuals have additional selection criteria for sites to come ashore at and tighter criteria for sites where they will pup and/or breed.

A distinction is made between haul-out sites and a colony ('rookery'). At haul-outs, seals may rest and socialise but pup births are rare. Colonies possess pups that can survive to weaning and may be self-sustaining units within a broader population.

Haul-outs are located across the foraging range of each species, and are usually in close proximity to foraging areas where there is minimal human disturbance. Such locations include offshore islands or rocky promontories backed by cliffs. They are places where seals can rest, conserve energy and avoid marine predators. In southern Australia, males and juveniles predominate at haul-outs, particularly those distant from core breeding areas. Males and juveniles are not compelled to return to colonies to support pups so can roam further than lactating females. They also tend to be the first settlers of new sites. Steps to colonisation include occasional visitation, regular visitation, increased numbers and then year-round occupation. Adult females may start to visit established haul-outs and pups may occasionally be born. Few haul-outs progress into colonies. To do this they require distinct features, including proximity to prey resources, seclusion from potential predators (notably man) and habitat qualities that improve pup survival (training pools, cover, wave-surge protection, easy water access and egress). Strongly influencing an individual's selection of a breeding site is natal site fidelity, access to breeding partners and previous breeding success. Simply having large numbers of seals present can be a strong advertisement to an arriving seal that a site is good. Steps to colonisation are occasional pup births, regular births with pups surviving to weaning and recruitment from within the breeding sites so that the colony becomes self-sustaining.

Life history

Most otariid seals share similar life history characteristics, including annual, synchronous breeding in spring and summer, sexual maturity at three to six years of age and longevity of approximately 18–20 years (Table 5.1). In Australian fur seals, New Zealand fur seals and Australian sea lions, maximum recorded longevity in the wild is 21, 23.4 and 26 years for females, and 19, 16.7 and 21.5 for males, respectively. Mortality is greatest (up to 70%) in the first year of life, then declines exponentially, with fewer than 1% of individuals achieving ages beyond 18 years. Death results mostly from starvation, although predators such as large sharks and Killer Whales take their share. Human-induced causes of mortality include entanglement in marine debris and drowning in fishing equipment. Along with demography and life-history traits, constraints on population growth are similar among the otariids.

The Australian sea lion is unique in being the only otariid that has a non-annual breeding cycle. Moreover, the breeding cycle is temporally asynchronous across the sea lions' range (i.e. adjacent colonies can breed at different times). Australian sea lions have the longest gestation period of any pinniped, protracted breeding and lactation periods, greatly reduced dispersal capacity, and extreme female philopatry (a tendency to return to their birthplace to breed) relative to other otariids. The evolutionary determinants of this atypical life history remain enigmatic. Key differences in reproductive biology are discussed in the following sections.

Territory establishment and maintenance

Most otariids exhibit a 'resource defence polygyny' mating system. Single males establish and defend territories where multiple females gather to pup. The piece of land, rather than the females themselves, is the resource defended by the males. Males may even establish intertidal territories and remain in them during high tides, ducking under as the waves wash over, while the females resort to higher ground. Territories of Australian and New Zealand fur seals typically contain five to nine (but up to 16) females during the breeding period. In Australian fur seals, average territory size is less than 60 m^2.

Table 5.1. Comparative life cycles for Australian fur seals and Australian sea lions.

	Australian fur seal		Australian sea lion	
	Female	Male	Female	Male
Pup	< 1	< 1	< 1.5	< 1.5
Juvenile	1 to 3	1 to 3	1.5 to 4.5	1.5 to 4.5
Subadult		3 to 6		4.5 to 7.5
Adult	> 3	> 6	> 4.5	> 7.5
(territorial)		(9 to 12)		(9 to 12)

Based on long-term marking and observation studies by Bob Warneke at Seal Rocks, male Australian fur seals mature at 3–6 years of age. but do not attain sufficient size to hold a territory until they are 8–13 years old. Then their breeding careers average just 2 years (maximum 6) before being terminated by younger and stronger individuals. The potential territory-holding males are termed bulls. Males that are too young, too old or otherwise incapable of establishing a breeding territory are evicted from the breeding areas during the breeding period. These may gather to rest in non-breeding areas, referred to as 'bachelor parks'. Large non-territorial males patrol the near-shore waters where they try to intercept and mate with departing females. They also make occasional sorties into breeding areas, attempting to secure territories from weaker territory holders.

In otariids that occupy breeding sites year-round, breeding-age males return to the colony throughout the year. Sally Troy was the first to record this behaviour in marked males during studies of New Zealand fur seals at Cape Gantheaume on Kangaroo Island, and the behaviour was confirmed through satellite tracking studies of males by Brad Page (SARDI Aquatic Sciences), at the same site. The males rest within the territories that they aim to secure during the breeding period, and actively evict other males from them. Troy related this behaviour to the males advertising their presence to both potential challengers and females that will seek secure territories during the breeding period. Occasionally, mating is observed outside the generally synchronised breeding period. Out-of-season mating often involves smaller sized females (potentially mating for the first time) and males that are not necessarily of the size that could hold a breeding territory.

As part of the breeding cycle of otariids, adults tend to go on extended foraging trips to gain energy reserves prior to the breeding season. The synchronised breeding cycle of fur seals involves territorial establishment by bulls in spring and then territory occupation for 30– 50 days (up to an extreme of 70 days). The timing of breeding events is slightly earlier in Australian fur seals than New Zealand fur seals. For example, territory establishment starts in late October for Australian fur seal bulls and late November for New Zealand fur seal bulls. Earliest arriving and biggest males tend to occupy the better breeding territories, which are around the high-tide mark where they can cool off if required, and on flat terrain. Carolyn Stewardson studied male reproduction in Cape fur seals and determined that males were capable of producing sperm between July and February, with spermatogenesis commencing 3–4 months prior to the breeding season (November–December), probably in response to increased photoperiod.

Territorial defence behaviours include vocal threats, huffing and snorting, ritualised posturing, bluff charges, snaps and bites, and rare fierce battles. Most ferocious battles occur during territory establishment (Plate 8a). Usually, fights last only a few minutes but they may last longer than 15 minutes. Savage wounds can be inflicted and wounds received can be life threatening. Contests end when one combatant flees, or vocalises submissively and backs away. Excessive effort in territory defence is avoided because the males fast throughout territory occupation and so need to conserve energy.

During very hot days, the territorial bulls utilise many strategies to avoid heat stress. They minimise movement, pant, seek cooler microhabitats within their territory, flipper wave, hold their flippers in water and, as a last resort, temporarily leave the territory for a cooling swim. Even mating appears arduous for territory-holding males on a hot day. Individuals may try to ignore females that attempt to initiate mating. Mating mostly occurs during the cooler hours of the day and night.

In Australian sea lions, a very long pupping season (4–9 months) means that female births are not well synchronised and males cannot monopolise mating opportunities to the extent that males of other species can. It simply is not possible for them to fast for the 120-day or more duration of the breeding season. Leslie Higgins, in her seminal studies on Australian sea lion behaviour in the late 1980s, described the mating tactics of males as a combination of resource and female defence. In contrast to other otariids, Australian sea lion males 'mate-guard' individual females, usually from when they haul-out to give birth through to their post-partum oestrus (0–10 days). Males may then seek another pre-oestrus female to 'mate-guard'. Higgins described this mating system as 'sequential polygyny'. Given the length of the breeding season, this mating tactic enables males to forage and build up energy reserves in between mate-guarding attempts. Further, with the breeding season being asynchronous among many colonies, males may try their luck at accessing oestrus females at several colonies within a year. Sequential polygyny obviously works for Australian sea lions but it has several drawbacks, such as requiring males to have an extended period of spermatogenesis, relative to other species.

Parturition (birth) and breeding

Heavily pregnant females (cows) start to arrive at colonies shortly after the prime territories have been claimed by the bulls. Each female will give birth to a single pup within 48 hours of arrival. In Australian fur seals, parturition takes place over a 5-week period between early November and mid-December, peaking usually in late November/early December. The date varies slightly between years. John Gibbens (PhD, University of Melbourne) found that in Australian fur seals, years of better female body condition in winter coincided with earlier than average mean dates of parturition and more births. All correlated with sea surface temperatures, seen as a predictor for prey abundance/availability.

In New Zealand fur seals, 90% of pups are born in a 5-week period between late November and early January. On Kangaroo Island, the breeding season peaks around 25–26 December. This is about 10–12 days later than for New Zealand fur seals on the South Island of New Zealand.

Peter Shaughnessy, Simon Goldsworthy and colleagues calculated the inter-pupping intervals of Australian sea lions using the intervals between 17 successive peaks in pup counts at Seal Bay (Kangaroo Island). The mean interval between the pupping seasons average 532 days or 17.5 months, but the range between breeding intervals was large, varying between 16 and 20 months. The pupping season is much longer in Australian sea lions, with 90% of births occurring over about 128 days, or 4.9 months, but the period over which all births occur can last up to 10 months.

Pups are most frequently born in the head-first presentation (Plate 8b), although breech (tail-first) births are common. The amniotic sac is usually broken during birth and the female often assists its removal from the pup. At colonies in southern Australia, the placenta is quickly detected and cleaned up by birds – mostly silver, kelp and pacific gulls. Hence during the breeding period, avid flocking and calling of gulls is a reliable sign of a pup birth. As for most mammals, aborted foetuses, premature births and still-born births all occur. Aborted foetuses may be seen at colonies from about June. Their frequency increases as the breeding period approaches. Premature births usually occur within 1 month of the breeding period. In one unusual case, however, a newborn Australian fur seal pup was present at Seal Rocks (Victoria) in late August. Females appear to mourn dead foetuses or still-born pups, and defend them against gulls and passing seals. They can sometimes be seen trying to stimulate and carry them, and show great reluctance to leave them.

Female otariids usually come into oestrus 5–10 days after giving birth. Males will keep track of the receptive state of females in their territory, occasionally sniffing them and the ground on which they sit. Copulation typically involves the male sitting over the female and to some extent restraining her. It can take place both on land (the majority) or in the water. Each copulation last for 10–30 minutes. After recovering from mating, females usually depart the colony to forage at sea. As one leaves the colony, she may be intercepted and mated by a patrolling bachelor male.

Gestation

Gestation in pinnipeds is characterised by a period of embryonic diapause (delayed implantation) following fertilisation. During diapause, the development of the embryo is arrested or greatly reduced. Embryonic diapause is believed to have evolved to enable seals to breed at regular intervals and give birth during favourable times of the year. In otariids, durations of diapause (3–4 months) and active (placental) gestation (8–9 months) enable the birth of the pup to occur 12 months after mating. Thus the pupping and mating can occur at the same time of year. In the walrus, *Odobenus rosmarus*, the period between births can be extended by lengthening the period between parturition and mating. In phocid seals, the timing of parturition is determined by a combination of delayed oestrus and extended embryonic diapause.

When the 17–18-month inter-birth interval of Australian sea lions was first noticed, it was thought to be achieved by extending embryonic diapause from 3–4 months to around 9 months. Adjusting the diapause rather than gestation period seemed the most parsimonious option, from an evolutionary perspective. Nick Gales and colleagues then examined oestradiol and progesterone concentrations in the females and discovered it actually was the gestation period that was lengthened. It was up to 14 months long or about 60% greater than in other seals.

Just how this species managed to break from the annual breeding cycle, and what environmental factors forced selection of an extended placental gestation (from 8–9 months to 13–14 months) has been the subject of much speculation. Despite the extended gestation, the Australian sea lion pups are not more developed at birth than those of other pinniped species. Thus, foetal development is probably slower than in other otariid seals. Slow foetal development could provide an energetic advantage by spreading the cost of gestation over a longer period. It may also allow females to direct greater energetic resources toward lactation and hence the growth of unweaned young. In an environment of highly unpredictable and variable prey availability, such a strategy may afford an energetic buffer to smooth out some of the environmental (prey availability) 'bumps'.

Female otariids maximise reproductive output, by suckling one pup while being pregnant with the next, but failed pregnancies reduce their success rate. For

Australian fur seals John Gibbens and colleagues recorded pregnancy rates of 70–90% at mid-term, then birth rates of 45–65%. These data suggest Australian fur seals have a high rate of foetal mortality during later stages of gestation. Both John Gibbens with Australian fur seals and Jane McKenzie (PhD, La Trobe University) with New Zealand fur seals found that mid-aged females (8–13 years of age) were better able to sustain a pregnancy through to parturition than were younger or older females.

Maternal attendance

In all otariid species, females remain with their pup for 5–10 days post-partum, after which they alternate between foraging trips to sea and attendance bouts ashore to nurse their pup and rest. This cycling termed 'maternal attendance behaviour' continues until pups are weaned (Figure 5.1). The bond between mother and pup is maintained by mutual recognition of calls and odour.

John Arnould and Mark Hindell found that Australian fur seal foraging trips could range from under 1 to more than 21 days. Trip lengths vary considerably within and between individuals and over the lactation period, from 3- or 4-day trips in summer to 6- or 7-day trips in winter. Attendance durations range from under 1 to over 8 days (averaging approximately 1.7 days) and remain relatively constant throughout lactation.

Simon Goldsworthy spent 2 years monitoring the maternal attendance behaviour of New Zealand fur seals at Cape Gantheaume, Kangaroo Island. He found that shore attendance bouts lasted about 1.7 days in duration and remained constant throughout lactation. Foraging trips, however, increased from 3–5 days early in lactation to 8–11 days late in lactation. However, foraging trips lasting more than 20 days were not uncommon. As a consequence of the increasing foraging trip durations, the percentage of time that mothers spend ashore decreased over lactation from about 24 to 14%. He found that pups compensate for the reduced availability of their mother by nursing more frequently and for longer periods when their mother is available. In so doing, they maintain an average nursing-rate of approximately 1.4 hours a day. The increasing durations of foraging trips occurs in response to the increasing energy demands of the pup and of gestation, and perhaps the increased costs of foraging over winter months when prey availability may be limiting or require greater travel time to access. Other factors influencing foraging trip durations include the fasting capacity of the pup, the energetics and foraging ability of the female and climatic/oceanographic variables.

Foraging trip durations of Australian sea lion females are short compared to those of other otariids. At Seal Bay and Dangerous Reef, studies by Leslie Higgins and

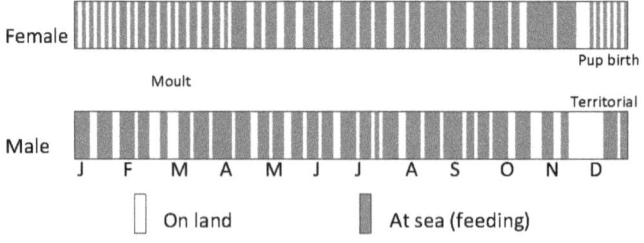

Figure 5.1. Stylised pattern of attendance ashore and feeding trips to sea for male and female Australian fur seals. Seals of both sexes will rest at sites other than where they breed.

Andrew Lowther, respectively, found foraging trips averaged 1.8 and 1.9 days. Foraging trips tended to be shorter during the first month, then increased and remained simular in duration until pups are weaned. Although foraging trip durations were similar at the two sites, periods ashore differed. Shore attendance bouts averaged 1.6 days at Seal Bay and 0.9 days at Dangerous Reef. The percentage of time the sea lions spent ashore averaged 48% and 32%, at the respective sites. The difference in colony attendance resulted from Dangerous Reef females also hauling-out to rest at sites away from their pups, while Seal Bay females did not. Dangerous Reef females could do this because they had appropriate nearby rest sites, whereas Seal Bay females did not. Including time spent hauled-out at other sites, Dangerous Reef and Seal Bay females actually spent similar proportions of time ashore (51%).

Lactation

In most mammals, if suckling is deferred for a period, lactation ceases and the mammary glands enter involution. This involves slowing milk secretion into storage alveoli in the mammary gland, a step that may be reversed if suckling recommences within 2 days. On cessation of milk secretion, there is an irreversible degradation of basement membranes and alveoli, returning the mammary gland to a rest state. How otariid females can sustain lactation through frequent and extended breaks (20 days or more) is curious and not completely understood. Obviously, the second stage of involution is somehow inhibited.

An otariid female normally suckles only her own pup (Figure 5.2). Should another pup approach her and attempt to obtain milk, it may be bitten and roughly flung away. As a pup grows, it aggressively deters other pups that seek milk from its mother. In some instances, though, female otariids will suckle a seal that is not their pup of the year. This may be deliberate support to a previous year's pup, or inadvertent or deliberate (alloparenting) suckling of an opportunistic pup or juvenile that is not related. The females may suckle other seals if their pup is not brought to term or dies prematurely. In Australian fur seals, because 45–65% of adult females give birth and 80–90% of adult females are lactating, there are many females available to suckle seals other than their pup of the year. Fiona Hume (Tasmanian Parks and Wildlife) and colleagues recorded that, at several colonies and throughout the day, over 20% of female Australian fur seals were suckling juvenile seals (between 1 and 2 years old). Possible sibling rivalry has been observed, with juveniles evicting pups at the teat, and on one occasion a female was observed simultaneously suckling a pup and a juvenile. Some pups and juveniles ('sneaky suckers') become adept at sneaking milk from more than one cow and can get exceedingly chubby. Others are poor at sneaky sucking and wake or irritate the females, so are quickly expelled. In the conspecific Cape fur seal, Steve Kirkman (South African Department of Environmental Affairs) recorded a bizarre case of a lactating female suckling an adult male.

Seal milk is richer in lipid (fat, up to 50%) and protein (10–18%) than the milk of terrestrial mammals, and can be transferred quickly to the pup. John Arnould and Mark Hindell noted that Australian fur seal milk averaged 42% lipid (fat) and 10% protein. Milk composition varied throughout lactation, with lipid levels increasing from 30 to 50% from early to late lactation, before decreasing to 45% prior to weaning (Figure 5.3). Protein content increased only slightly during lactation from 10 to 12%. In New Zealand fur seals, lipid levels are lower, averaging 37% (range 22–57%); but

Figure 5.2. Australian fur seal cow suckling its pup. Photographer: Roger Kirkwood.

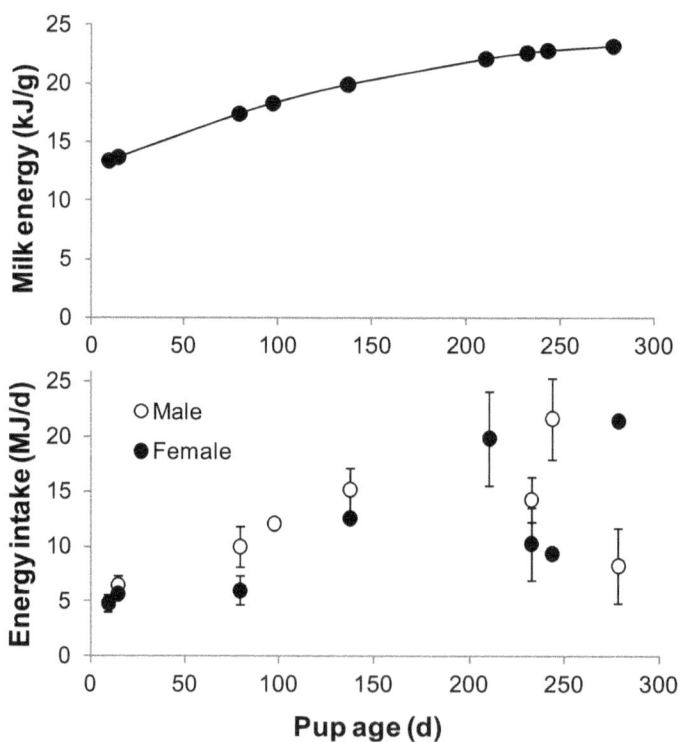

Figure 5.3. Change in relation to pup age in (a) the energy content of Australian fur seal milk and (b) the energy intake per day by the pup. Derived from Arnould and Hindell (2002). © John Wiley and Sons.

protein levels are slightly higher, 15–16% (range 11–20%). Average milk consumption by Australian fur seal pups was 557 g/day, ranging from 400 g/day shortly after birth to 675 g/day at 210 days of age.

Australian sea lions have relatively low lipid content in their milk. Maria Krezmann, Nick Gales and colleagues reported that the lipids increased from 26% in early lactation (under 250 days) to 39% in late lactation (over 250 days) and Andrew Lowther found a lipid level of only 21% in early lactation. The low lipid content could be related to the sea lions' very short foraging trips. In many otariids, there is a strong positive relationship between milk lipid levels and the duration of the preceding foraging trip. The capacity to increase the energy content of milk partially compensates pups for the longer fasting periods and enables females to facilitate the extension of foraging ranges during lactation.

Female otariids that rear pups year round (all those in southern Australia) essentially lactate for 8–10 months each year for their entire adult lives. This is typical for otariids dwelling in temperate latitudes. Those living in higher latitudes, such as Antarctic and northern fur seals, wean their pups after 4 months (so do not lactate for 8 months of each year). The shorter lactations at higher latitudes possibly relate to the colder climate affecting survival of pups ashore. At Macquarie Island, Iles Crozet and Marion Island in subantarctic waters, Antarctic and subantarctic fur seals breed alongside one another, and exhibit 4- and 10-month lactations, respectively. It is unclear which strategy is most optimal for this environment. One of the longest lactation periods of any otariid is the 15–18 months exhibited by female Australian sea lions. If a female fails to pup in consecutive seasons, it will generally nurse its pup for a further 15–18 months and occasionally over three breeding seasons (i.e. more than 4 years). For species like the Australian sea lion, the notion of a defined lactation period that corresponds with the weaning of the pup may be incorrect, and females essentially may be in a perpetual state of lactation. The extended period of lactation given to pups is certainly likely to aid their survival, but it reduces the potential lifetime reproductive output of the female.

Female otariids are clearly built to nurture. For much of their adult life, they are pregnant, lactating and so are feeding to support three lives (the pup *in utero*, the pup on the teat and themselves).

Pup growth to weaning

Male otariids are generally heavier than females at birth, grow faster and are heavier at weaning. Observations of pup behaviour suggest there is no difference in the periods of nursing of the sexes or quality of milk delivered. One theory as to why males could grow faster is that they may suck harder. Different growth of pups may also be influenced by the types of body tissues being laid down. Comparatively, males direct more energy towards lean tissue (muscle) development whereas females direct more to fat. Muscle is heavier than fat, so males may gain more mass than females for the same milk energy received. Growth rates of pups decline markedly as weaning is approached.

John Arnould and colleagues determined that the average birth mass of female and male Australian fur seal pups was 7.2 and 8.3 kg, respectively. Average growth rates from birth to weaning were 53 and 62 g/day (Figure 5.4). Pups doubled their birth mass after approximately 135 days, which was longer than in most other fur

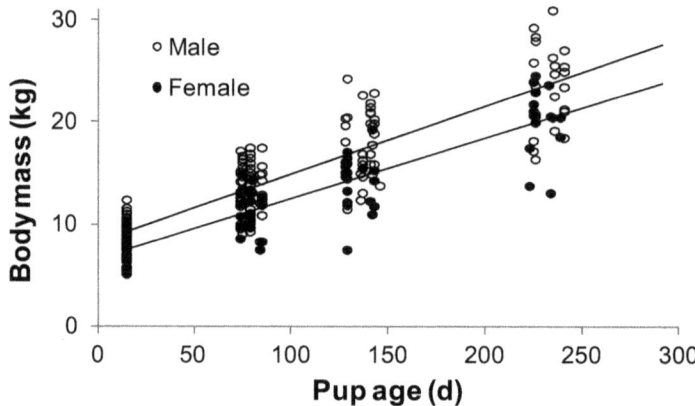

Figure 5.4. Growth of male and female Australian fur seal pups. Derived from Arnould and Hindell (2002). © John Wiley and Sons.

seals. Mass at weaning (about 300 days) was about 23 and 27 kg for females and males, respectively. Most pups weaned between September and October. At that time, adult females departed for extended foraging trips to prepare for their next birth.

New Zealand fur seal pups weigh between 3 and 4 kg at birth; males double their mass in 60–100 days and females in 80–90 days, and mean weaning weights are 13–16 kg. Tami Haase (PhD, La Trobe University) studied the determinants of weaning in New Zealand fur seals. She found that some pups weaned as early as August and others as late as mid-December, but 90% weaned over a 57–67-day period, mid-September to late November. Mean weaning age was 9.8 months. Although the median pupping date (when half the pups have been born) did not vary between the 3 years investigated, the median weaning dates did, resulting in a 10-day range for mean age at weaning. Haase determined that pups whose mothers spent a greater proportion of time ashore (made relatively shorter foraging trips and had longer shore bouts), and who attained greater weight in early growth, tended to wean at a younger age. Pups weaned themselves without any apparent conflict with their mothers, who continued to return to the colony after their pup had weaned.

Australian sea lion pups weigh 7–11 kg at birth. Growth rates over the first 100 days can be colony-specific; Leslie Higgins recorded 122 g/day at Seal Bay and Andy Lowther 89 g/day at Dangerous Reef. Higgins monitored nine pups to weaning and noticed growth rates averaged 84 g/day over the first 12 months, then 43 g/day between 12 and 18 months. Australian sea lion pups weigh 30–40 kg at weaning.

Pup transition to independence

In seals, the transition from dependence to independence is associated with high rates of mortality. This transition represents a unique challenge because there is both spatial and temporal separation between the colony where pups are nursed and the sea where their mothers forage. Pups improve their chance of survival after weaning by attaining the greatest possible weight and by learning to catch prey early, even doing so while still being dependant. When mothers are at sea, pups hone their swimming, diving and foraging skills: much of the in-water 'play' of seal pups is directed at developing these skills. Even so, seal pups are often naive of their prey when they wean. The degree of naivety varies markedly between species. Generally, it is greatest in species with short lactation periods where weaning is abrupt and least in species with prolonged lactation where weaning is gradual.

From July, Australian fur seal pups start supplementing their milk diet with self-caught fish. Up to this time, they remain within 50 m of the shore because pups that drift away from colonies prior to July will not be located by their mothers and are likely to die. After July, some pups start to venture further out and some may achieve independence as early as August. Lisa Spence-Bailey found that the oxygen-carrying capacity of the blood of Australian fur seal pups steadily increased through time, reaching 71% of adult levels prior to weaning. Pups increased their time in the water from under 8% in the first months to about 30% after moult. Just prior to weaning, they could dive to over 60 m, although dive durations were short, typically under 1 minute.

Alistair Baylis (Honours project at La Trobe University) investigated the development of foraging skills in New Zealand fur seal pups at Cape Gantheaume, Kangaroo Island. Between June and September, pups dived successively deeper (6–44 m). The average number of dives per day, dives per hour and vertical distance travelled also increased. Some pups were catching prey in June, when 5–6 months old, as indicated by prey remains in 30% of pup scats.

Australian sea lion pups perhaps develop perhaps the most proficient foraging skills prior to weaning of any otariid. Up until about 4 months of age their excursions are limited to rock pools and near shore waters: when 6 months old, their diving and foraging behaviour develops in earnest. Shannon Fowler (PhD, University of Santa Cruz) at Seal Bay, Kangaroo Island, noted that 5–7- and 13–16-month-old pups dived to mean maximum depths of 29 and 68 m, respectively. The mean maximum distances travelled by the 13–16-month-old pups ranged between 16 and 33 km. Andrew Lowther's PhD work (University of Adelaide) in the Nuyts Archipelago confirmed that from 6 months of age pups were proficient divers and by weaning age (17.5 months) were essentially undertaking foraging trips and shore bouts in a similar fashion to adults. In Australian sea lions, pups have a long period over which foraging skill can be developed while retaining the safety net of maternal lactation. This 'soft landing' transition to independence undoubtedly increases the chances of post-weaning survival in this species.

The unusual breeding pattern of Australian sea lions

A number of studies have identified the unusual non-annual (17.5 months) and temporally asynchronous breeding pattern in Australia sea lions (Figure 5.5). Nick Gales and colleagues noted there seemed to be no geographic (proximity) pattern that explained the level of breeding asynchrony among colonies. The timing, causes and benefits of this unusual breeding biology have intrigued many seal biologists.

In most otariids, linking the breeding strategy to an annual cycle allows pups to be born at a time of year when their chances of survival ashore are optimised. This invariably is in spring and summer. Furthermore, linking to an annual cycle drives a strong breeding synchrony and individuals can maximise their condition for a finite period when breeding occurs. To break the link with the annual cycle suggests both offspring

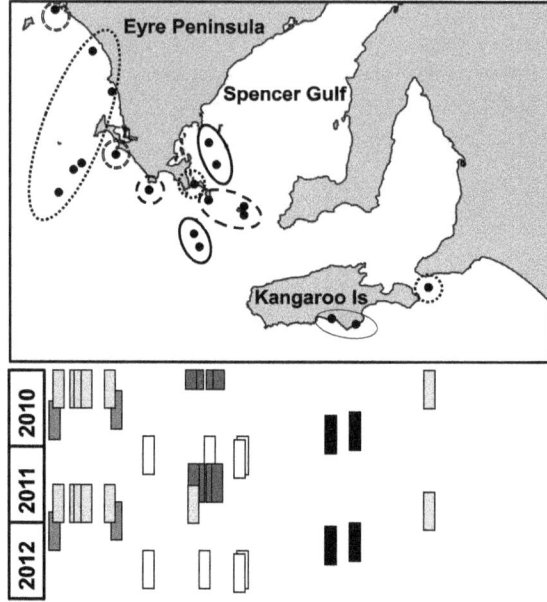

Figure 5.5 Australian sea lion breeding chronology at selected colonies in South Australia. The map indicates colony locations; those with similar breeding chronologies are circled similarly. The time-chart below for years 2010 to 2012 contains grey-scale blocks that correspond with locations on the map and indicate the 6-month pupping periods at those locations (shading links synchronised colonies). This figure depicts the asynchrony of breeding among locations and the unique 1.5 year breeding cycle of Australian sea lions. Simon Goldsworthy, unpublished data.

survival and reproductive success are not enhanced by a synchronised, spring–summer birth and breeding period.

In 1997, Nick Gales and Dan Costa suggested that a major determinant of the breeding strategy of the Australian sea lions could be their nutrient poor, stable marine environment, where seasonality of breeding offers no advantage to the temporal patterning of energetically expensive lactation. By extending the period of lactation and care of young, Australian sea lions may spread their effort over a longer period, during which dependent young can supplement their nutritional requirements and learn foraging skills that are necessary for opportunistic feeding. However, Gales and Costa acknowledged a contradiction with their hypothesis. The annually breeding Australian and New Zealand fur seals occurred in sympatry with Australian sea lions. More recently, oceanographic research by John Middleton (SARDI) and others has demonstrated that shelf waters off southern Australia are not consistently stable and nutrient poor. Indeed, waters off South Australia where more than 80% of the population of Australian sea lions resides are highly seasonal and amongst the most productive across southern Australia.

Simon Goldsworthy proposed an alternative explanation for the unusual breeding strategy of Australian sea lions which incorporates a 'family farm'-like inheritance of maternal foraging space and behaviour. The premise for the 'family farm' hypothesis is built upon the species's extreme female philopatry and the marked within – and between – colony differences in feeding behaviours of adult females, more so than in other otariids. These suggest there is strong evolutionary selection for females to stay within their natal home range and adopt unique and specialised foraging strategies. In the same manner as 'family farm' inheritance in human societies, foraging specialisations, including foraging habitats and prey species, could be passed down by sea lions from generation to generation. Extending the period of maternal care would enhance the capacity for social transmission (learning) of foraging behaviours. However, Andrew Lowther and colleagues found no evidence that females that shared particular specialised foraging strategies (termed ecotypes) also shared the same maternal (mitochondrial DNA, mtDNA) lineages. This may indicate that instead of vertical (between generation) social transmission, foraging behaviours may be learned horizontally, by within generation social learning. Alternately, social learning of foraging behaviours may occur through a mixture of vertical and horizontal means, or that maternal lineages of foraging are maintained only during periods of stable environmental conditions and are regularly disrupted by environmental change that occurs over time lines too short to leave a genetic signal.

Unlike most other otariids, Australian sea lion pups are believed to undertake at-sea excursions with their mothers, with 4–5-month-old pups regularly being observed nursing from their mother at islands 20–60 km from their natal colony. It is unlikely that pups make such long journeys on their own. While following its mother at sea, an Australian sea lion pup may acquire fine-scale knowledge of its marine environment, including where its mother forages, what prey she feeds on and how she captures it. Pups may acquire such information from watching other sea lions, too.

This social learning of foraging skills and behaviours may occur in pups of either sex, but genetic data indicate that only females remain highly philopatric throughout their lives. Males ultimately adopt a very different foraging pattern to that of their mothers, principally driven by the different reproductive strategies of the two sexes. How the reproductive strategies differ is described more fully in the following chapter on foraging ecology.

The Australian sea lion's reproductive strategy can be seen as a novel approach to maximising reproductive success following departure from an annual breeding paradigm. It represents a unique way to deal with a seasonally productive but variable environment, achieving reduced daily energy needs, greater investment in individual offspring and reduced costs to residual fitness (future reproductive potential).

6
FORAGING ECOLOGY

The foraging strategies of individual seals evolve through life as a series of trade-offs between maximising energy gain and reproductive output, and minimising risk. They vary depending on the availability of prey, which changes temporally and spatially, and the age, experience and body condition of the individual seal. Fish, cephalopods, crustaceans, seabirds and occasionally other seals can contribute to the diet of otariid seals. The prey generally exhibit patchy distributions with concentrations around oceanographic or benthic (bottom) features, such as current boundaries, upwelling zones, seamounts and reefs. Seals forage in different zones within the ocean (Figure 6.1). Fur seals tend to be pelagic (open water) foragers that target mobile prey that occur in large, patchily distributed schools. In contrast, sea lions tend to be benthic foragers that target both benthic prey, whose availability changes over time through settlement, growth or temporary emergence from a cryptic habitat, and demersal (at or near the bottom) prey that school in association with the sea floor. Benthic and demersal prey also will have a patchy distribution. While pelagic foragers often source prey

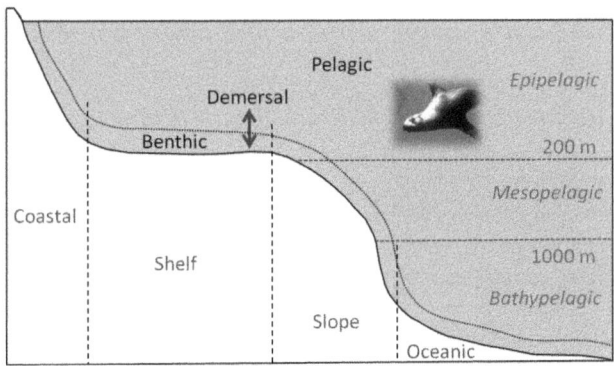

Figure 6.1. Stylised zonation of the ocean indicating benthic (associated with the bottom), demersal (migrate on and off the bottom) and pelagic (open water) foraging habitats utilised by seals.

beyond the continental shelf, benthic foragers are constrained to shallower coastal and shelf regions due to the physiological limits of dive depth and duration.

Otariid seals tend to forage out of a central place which may be fixed or, if the individual migrates between regions, may change seasonally. Out of each place, the individual undertakes feeding trips that vary in distance and duration. Of critical importance to the selection of 'central places' and to durations of occupation is proximity and consistency of food supply. Some seals are prone to focus on 'hot spots' while others roam more widely.

One critical central place for a mature otariid is its breeding site. Most otariids display some level of fidelity to a breeding site, which might be visited year-round or only during breeding periods. Female otariids have a particular attachment to their breeding site because they need to return there semi-regularly during their 4-or-more month lactation periods, to support their dependant pup. Prey resources must be encountered consistently within striking distance of the breeding site for pup rearing to be successful. Because foraging habitat quality is not the only reason influencing breeding site selection, the quality of foraging habitats adjacent to breeding sites will vary.

Techniques

Knowing what a seal eats is important to understanding its foraging ecology and potential trophic (food web) interactions. Researchers often assess seal diet from prey hard parts that remain undigested in regurgitates and faecal samples (Figure 6.2). The most useful hard parts are fish otoliths (ear bones) and cephalopod beaks. These have species-specific sizes and shapes and can allow estimation of the size and age of the prey. The prey revealed from regurgitates can differ from those revealed from faeces due to the small size items need to be to pass through the digestive tract, as well as erosion of items during passage. Consequently, larger bodied prey items are often better represented in regurgitates.

Prior to scat analysis, seal diets were determined by observation of predation or by killing an animal and examining its stomach contents. Scat analysis provides a better alternative. There are biases in scat analysis, including poor representation of prey that have no hard parts or whose hard parts are not eaten. One method to detect such softer bodied food items is through identification of prey DNA in scats. Clues to diet can also come from an analysis of stable isotope ratios. A higher level of the carbon isotope ^{13}C suggests foraging is being conducted closer to shore, and higher levels of the nitrogen isotope ^{15}N suggest predation occurring at a higher trophic level. Different tissues of the seal can provide clues to diet over different time scales. For example, blood plasma levels reflect the last meal, red blood cell levels reflect the past week of feeding, fat levels indicate weeks to months and levels in sections of whiskers can allow diet to be investigated across a year or more. Stable isotope levels in museum specimens (including teeth and bone, and soft tissue) can allow relative diets to be investigated over decades or more.

Before the 1980s, the foraging habits of otariids were largely concealed beneath the waves. There were some insights from dietary examinations, tagging studies and opportunistic observations. Thereafter, electronic instruments attached to seals began to reveal locations at sea, dive patterns, physiological conditions of the individuals and environmental variables, such as water temperature (Figure 6.3).

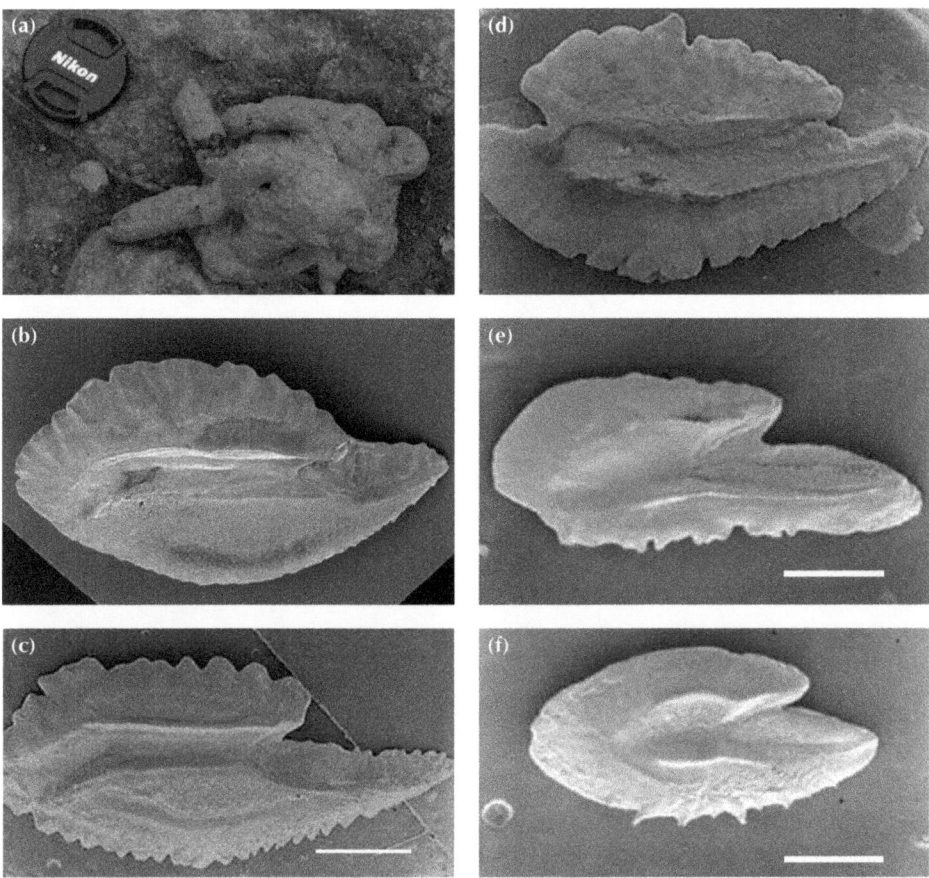

Figure 6.2. (a) Australian seal scat sample and otoliths (ear bones) of fish that can be sieved out of a seal scat; (b) jack mackerel *Trachurus declivis*; (c) redbait *Emmelichthys nitidus*; (d) barracouta *Thyrsites atun*; (e) sardine *Sardinops sagex neopilchardus*; and (f) anchovy *Engraulis australis*. Photographers: (a) Roger Kirkwood, (b–f) Simon Goldsworthy.

Concomitant with instrument developments has been refinements of capture and restraint procedures, and means of attachment, such as marine-stable glues. Devices are generally glued to the dorsal fur of seals and, if not recovered, will fall off the individual during its next moult.

The main technological advance enabling scientists to collect data on otariids at sea was the development and miniaturisation of archival tags resistant to marine conditions. Investigations during the 1960s, '70s and '80s led to widespread adoption of time–depth recorders in the 1990s to supply two-dimensional views of seal dives. Coinciding with dive recorder developments were developments in animal tracking.

One procedure to track an animal is to use VHF transmitters, receiving stations and triangulation. This is difficult and labour intensive to do for far travelling marine animals. In 1979, animal tracking was revolutionised by the Argos satellite-based system, a collaboration between the French Space Agency (CNES), the US National

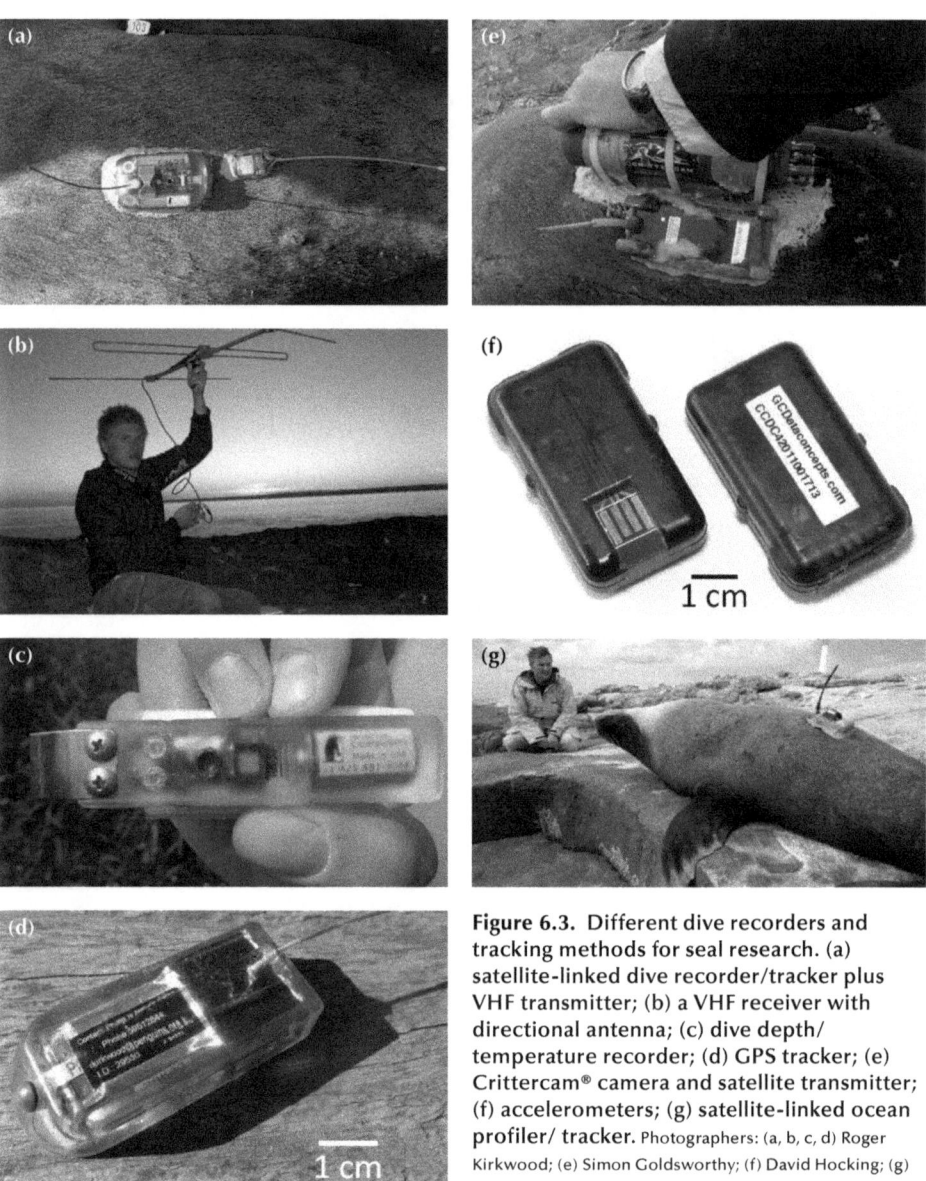

Figure 6.3. Different dive recorders and tracking methods for seal research. (a) satellite-linked dive recorder/tracker plus VHF transmitter; (b) a VHF receiver with directional antenna; (c) dive depth/ temperature recorder; (d) GPS tracker; (e) Crittercam® camera and satellite transmitter; (f) accelerometers; (g) satellite-linked ocean profiler/ tracker. Photographers: (a, b, c, d) Roger Kirkwood; (e) Simon Goldsworthy; (f) David Hocking; (g) Roy Hunt.

Oceanic and Atmospheric Administration (NOAA) and the US National Aeronautics Space Administration (NASA). By the late 1980s and early 1990s, platform transmitter terminals were sufficiently small, sturdy and reliable for deployments on seals. Each platform transmitter terminal transmits a unique code that can be detected by polar orbiting satellites and forwarded to ground processing centres. Platform transmitter terminal locations are then predicted by measuring the Doppler shift on the transmitted signal that is received during the satellite pass. The accuracy of the estimated locations varies depending on how many transmissions per pass the satellite detects from the platform transmitter terminal, the angle overhead that the satellite passes and transmitter frequency stability. The highest quality locations provide an accuracy of better than 150 m. Data usually are filtered based on location quality and either maximum likely speeds of the animal or interpretations of changes in the animal's behaviour over time (e.g. state–space modelling). Packaged deployments of platform transmitter terminals and time–depth recorders on the same seal allow three-dimensional interpretations of animal movement.

During the early 2000s, Global Positioning System (GPS) tags were developed for marine animal tracking. Unlike platform transmitter terminals that transmit data to satellites, GPSs tags receive satellite transmissions and process them on board to estimate locations. They may then transmit the locations to a satellite or ground receiver, or store them until the animal is recaptured. GPS devices provide more locations and greater accuracy than do platform transmitter terminals.

Numerous other devices have been developed and deployed on seals to learn more of their three-dimensional travel and feeding behaviour. Examples are geolocation (GLS) tags that record light to enable tracking based on times of sunrise and sunset; accelerometers that record micro-bursts in speed and changes in direction; stomach temperature or mouth-opening recorders to investigate moments of prey ingestion; gyroscopic pitch, roll and yaw recorders; heart rate recorders to investigate energetics; and still and video cameras. One camera example is the Crittercam®, which was developed with support from National Geographic and first deployed on otariids in 1992. Improvements in data and power storage have expanded the information obtainable from such instruments, and reductions in size have increased the range of species on which they can be deployed.

Methods for seal capture and restraint of seals vary depending on the size of the seal, the terrain and the purpose for capture (Plate 9). Researchers need to take into account that capture procedures and carrying devices can cause an animal to change its behaviour. Such changes to behaviour have to be minimised to reduce adverse impacts on the individual and biases to the data. Animals that exhibit a strong fear of humans or easily go into shock are prone to respond badly to restraint. Amongst the three Australian otariids, Australian fur seals are particularly flighty (Figure 6.4),

Figure 6.4. Australian fur seals ashore are very alert to disturbances and will stampede to the water if they feel threatened. Photographer: Roger Kirkwood.

New Zealand fur seals are a little less flighty and Australian sea lions tend to be more nonchalant or confrontational towards approaching humans. Techniques for minimising the stresses of capture include minimising pursuit distances and restraint times, avoiding more nervous individuals, remote darting, utilising anaesthesia for periods of extended restraint (over 10 minutes) and, especially, sharing of knowledge between researchers and veterinarians. As an example of gained knowledge, phocid researchers tend to take blood from the pelvic vein in a post-pelvic, vertebral sinus, whereas otariid researchers have better success with the pectoral vein (Plate 10a).

Otariids rely on their hydrodynamic body shape and flexibility to minimise the energetic costs of transport and maximise prey capture. Any device attached to them adds drag and compromises swimming ability. The more hydrodynamic, light and small the device, the less it will interfere.

Attached devices have enabled greater resolution of potential conflicts and interactions between otariids and human activities, assisted conservation and management strategies, and provided a better understanding of the habitat features that underpin the sustainability of seal populations. It is essential, however, that researchers and ethics committees constantly weigh up the value of the data sought against the impacts of the research on individuals and populations.

Australian fur seal
Ranges
In the 1970s, Bob Warneke found that Australian fur seals flipper-tagged as pups on Seal Rocks could be recovered over 700 km away. Since the early 1990s, foraging ranges of individual seals from several colonies have been assessed using geolocation tags and platform transmitter terminals. Research has focused on lactating females, whose ranges are restricted as they need to return to the colony regularly to suckle their pup. This obligation coincidently facilitates the recapture of individuals to recover instruments. In 1997–99, John Arnould and Mark Hindell attached geolocation time–depth recorders to females at Kanowna Island. The females foraged principally in central Bass Strait but could range up to 300 km from their colony.

Researchers have since deployed platform transmitter terminals to lactating females at many Australian fur seal colonies. This included a study that simultaneously assessed 3 years of winter–spring foraging tracks by females at four Victorian colonies (Plate 10b). Females from The Skerries, situated on the narrow shelf of Australia's south-east coast, tended to travel shorter distances than did those from Lady Julia Percy Island, located in north-western Bass Strait on a slightly broader shelf to the west of Bass Strait. Most females from The Skerries foraged within 20–60 km of their home, whereas those from Lady Julia Percy Island tended to forage 20–150 km from their colony. Females at Seal Rocks and Kanowna, both situated centrally in northern Bass Strait, also tended to forage 20–150 km from their colonies. Seals from the two central-northern Bass Strait colonies had the largest foraging ranges.

While most research has been on female foraging, there are some data on males. In 1993, an adult male Australian fur seal that had frequently been trapped at a fish farm in southern Tasmania and released 450 km away in Bass Strait was fitted with a geolocation time–depth recorder. It was re-trapped 15 days later back at the fish farm. The seal had utilised shelf waters off eastern Tasmania and rested at several known seal haul-outs along its route. In 1999–2001, adult males from Seal Rocks were tracked (Plate 10b). Most foraged in western Bass Strait. Some also travelled down the west coast of Tasmania to forage in southern Tasmanian waters, 500 km from Seal Rocks, and one foraged west of the Eyre Peninsula (South Australia), 1200 km from Seal Rocks. Juvenile seals tracked from Lady Julia Percy and Seal Rocks display similar ranges to adult females, although male juveniles tend to be more adventurous than female juveniles and travel further afield.

Simon Goldsworthy and Roger Kirkwood investigated the foraging locations and ranges of males that were interacting with fishing trawlers along the west coast of Tasmania. These males were caught at sea using a large 'dip-net' that was lowered by the ship's crane (see Plate 9f). Tracked seals continually targeted the fishing operations, resting between foraging trips at haul-outs on Tasmania's west coast, until the fishing season ended. Then the seals moved to forage in southern Tasmania or Bass Strait.

Australian fur seals feed mostly in association with the sea floor and rarely leave the continental shelf, which reflects the benthic nature of their foraging. This foraging behaviour is more sea lion-like and contrasts with the mid-water (pelagic) foraging of other fur seals, including the conspecific Cape fur seal and sympatric New Zealand fur seal. However, Australian fur seals are capable of foraging pelagically. For instance, several tracked seals have ventured off the shelf, including one adult male that spent several weeks in water depths of over 1000 m to the west of the southern tip of Tasmania; the seal could not have foraged benthically in this area.

Foraging trip strategies

Australian fur seals depart from shore at all times of the day, although there is a peak in departures in the few hours leading up to sunrise. They immediately commence bottom diving and feed if they encounter prey. After a day or more of travel, the seal may arrive in a preferred foraging area and remain there for 5 days or more before returning to land, usually the site from which it departed. While some seals repeatedly return to uniquely preferred 'hot spots' within their ranges, others alternate between two or more focal areas and a few seals appear to 'roam' almost randomly during foraging trips. Routes can weave or traverse in circular patterns. Factors influencing changes in trip time and foraging area could be weather patterns, competition from other seals, encounters with predators, or encounters/non-encounters with prey concentrations. If foraging takes an individual some distance from the site it departed from, it will rest at alternative, closer sites; even lactating females do this.

Diving behaviour

Australian fur seals can continue to bottom-dive for up to 36 hours at a time. Most dives are U-shaped, which is indicative of direct descent, travel along a near-flat substrate (or waiting in ambush), then direct ascent. Footage from Crittercam® cameras attached to females at Kanowna Island by Andrew Hoskins and John Arnould (to be published) confirm that the seals track along the bottom on most dives. This behaviour is unusual amongst fur seals although it is common amongst benthic foraging sea lions. Mean dive durations of individuals range from 2 to 4 minutes, with durations up to 10 minutes being recorded. Most dives are to depths shallower than 80 m. The maximum depth recorded by a time–depth recorder has

Table 6.1. Foraging habitats and key prey of the three Australian otariids.

Habitat		Australian sea lion	Australian fur seal	New Zealand fur seal
Habitat		On shelf, local	On shelf, local and distant	On and off shelf, local and distant
		Principally benthic	Principally benthic/ demersal	Principally epipelagic
Prey		Fish, cephalopods, crustaceans	Fish, cephalopods	Fish, cephalopods, seabirds
Top 10	1	Octopus sp	Redbait	Redbait
	2	Calamari squid	Barracouta	Sth Ocean arrow squid
	3	Arrow squid	Jack mackerel	Arrow squid
	4	Giant cuttlefish	Gurnard	Calamari squid
	5	Sharks and rays	Red cod	Barracouta
	6	Southern rock lobster	Arrow squid	Jack mackerel
	7	Leatherjackets	Leatherjacket	Myctophids
	8	Australian salmon	Tiger flathead	Swallowtail
	9	Swallowtail	Silver trevally	Western gemfish
	10	Flathead	Calamari squid	Little penguin

been just over 160 m; however, cameras fitted to trawl-nets have recorded Australian fur seals at depths below 200 m.

Australian fur seals forage predominantly during daylight hours, although some individuals are mainly nocturnal foragers and others forage without any apparent regard for time of day. They forage almost exclusively across the continental shelf waters and appear to prefer depths of 60–80 m.

Diet

Australian fur seals are generalist predators with 60–70 species of fish and cephalopod species recorded in their diet. The importance of particular prey varies geographically, but the principal ones consumed include jack mackerel (*Trachurus declivis*), redbait (*Emmelichthys nitidus*), barracouta (*Thyrsites atun*), arrow squid (*Nototodarus gouldi*), leatherjackets (family Monacanthidae), gurnards (family Triglidae), red cod (*Pseudophycis bachus*) and flathead species (family Platycephalidae) (Table 6.1).

In the past 200 years, humans have modified the ecosystems in which Australian fur seals inhabit. First the seals were harvested to near extinction, then commercial fisheries reduced some prey and provided access to others (such as by bringing deepwater species to the surface in trawl nets). In the 1990s, the broad-scale elimination through an introduced herpes virus of the once ubiquitous sardine (*Sardinops sagax neopilchardus*) again influenced trophic structures across southern Australia. Although sardine stocks near the centre of the outbreak (in South Australia) recovered quickly, in Bass Strait 15 years after the events they had not recovered.

Several studies have investigated Australian fur seal diet. During the 1920s and 1940s at colonies in northern Bass Strait, Australian fur seals shot for research purposes contained in their stomachs remains of barracouta, cephalopods and southern

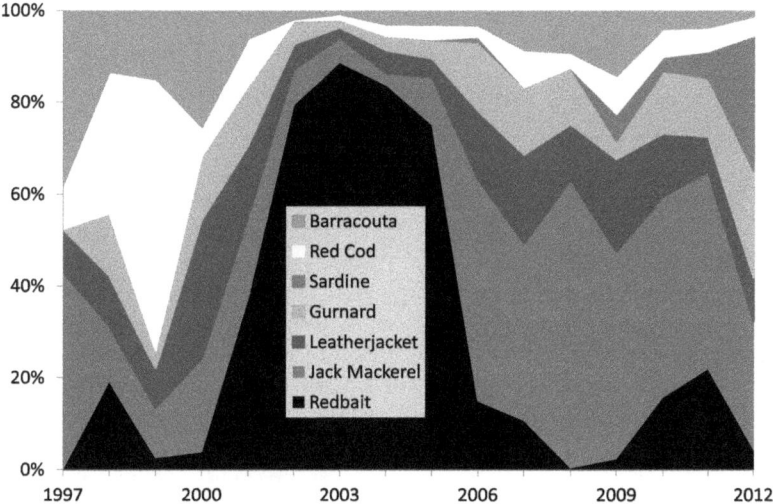

Figure 6.5. Summary by year of the diet of Australian fur seals at Seal Rocks, based on the proportional frequency of occurrence of fish remains in scats collected on a near-bimonthly frequency. Adapted from Kirkwood, Hume and Hindell (2008) and Roger Kirkwood unpublished.

rock lobster (*Jasus edwardsii*). In the 1980s and 1990s at sites around Tasmania, the diet determined from prey hard parts in scats and regurgitates comprised redbait, jack mackerel, leatherjackets (family Monacanthidae) and arrow squid (*Nototodarus gouldi*). During 2001–02 in South Australian waters, redbait was the dominant prey of adult male Australian fur seals. Also, between 1998 and 2012, a study by Roger Kirkwood and colleagues examined near-monthly scat samples collected from Seal Rocks, to examine multi-year changes in diet. Redbait were scarce for 4-year periods, then constituted 40–50% of fish prey recorded for 4-year periods. When scarce, redbait were replaced by jack mackerel, barracouta, leatherjackets or red cod (Figure 6.5). John Arnould examined stable isotope ratios in seal tissues at nearby Kanowna Island, and found a similar multi-year pattern of change; he suggested the redbait years correlated with years of poorer body condition. Such multi-year dietary shifts point to cyclic ecosystem changes in Bass Strait, possibly related to large-scale, oceanographic processes.

Much of the dietary assessment to date has relied on finding prey remains in scat. There are distinct biases with this technique, such as prey with non-resilient hard parts being poorly represented. Bruce Deagle (Australian Antarctic Division) and others investigated the presence of prey DNA in 90 scats from each of three colonies across northern Bass Strait. They successfully extracted DNA of 62 prey species and revealed a mean diet of 80% fish and 20% cephalopod, with minor amounts of cartilaginous prey and traces (or potential sample contamination) of crustacean and avian prey.

Human influences and cyclic oceanographic processes are likely to continue to influence ecosystem structure and prey availability for Australian fur seals. For example, human-influenced climate change is resulting in an overall increase in mean sea-surface temperature around southern Australia, with temperatures in the southeast predicted to increase substantially more than global averages. On the east coast of Tasmania, an increase of over 1°C was recorded between the 1940s and 2000s.

Modelling of impacts of climate change on marine communities predicts a range of possible scenarios for the short and long term.

New Zealand fur seal

Ranges

In one of the first seal tracking studies using platform transmitter terminals, Rob Harcourt found that female New Zealand fur seals from the Otago Peninsula, New Zealand, demonstrated seasonal changes in foraging range. They foraged near the shelf slope during summer, farther offshore in autumn, then in coastal waters during winter. The changes were assumed to reflect changes in prey abundance and availability over time. Subsequent satellite tracking studies in South Australia by Brad Page and Alastair Baylis also indicated seasonal changes in foraging range. There, coastal habitats are influenced by seasonal nutrient enrichment, driven by the Flinders Current. A body of Antarctic surface water lies at depth in waters south of Australia and intermittently wells up onto the shelf of south-west-facing coastlines. The Bonney Coast of eastern South Australia and western Victoria, western Kangaroo Island and the lower and western Eyre Peninsula are regions where upwelling of these cold, nutrient-rich waters occurs. Upwelling generally occurs between November and May, but is usually strongest in late summer (February–March). Much of the upwelling is subsurface and creates a strong thermocline usually between 30 and 60 m, where productivity is enhanced and zooplankton, fish and squid are often concentrated. It is these resources in the water column that New Zealand fur seals appear adept at targeting.

Brad Page compared the foraging behaviours of adult females, adult males and juveniles at Cape Gantheaume (Kangaroo Island), while Alastair Baylis compared the foraging patterns of adult females at four major breeding colonies: Cape Gantheaume, Cape du Couedic, Neptune and Liguanea islands. Page identified marked differences in the foraging regions used by juvenile, adult female and male seals. Juveniles forage in oceanic waters at great distance from the breeding colony (1095 km mean maximum). Lactating females foraged predominantly in mid to outer continental shelf waters (about 108 km mean maximum distance), whereas adult males foraged in generally deeper continental slope waters (188 km mean maximum distance).

Baylis noted inter-colony differences in female foraging ranges (Figure 6.6). In autumn months, most females from Cape du Couedic, North Neptune and Liguanea Island foraged in oceanic waters associated with the Subtropical Front, an interface between temperate and subantarctic surface waters that is situated 400–800 km south of their colonies. As with Page's study, Baylis found that most females at Cape Gantheaume foraged in mid to outer shelf waters, associated with the seasonal coastal upwelling off the Bonney Coast. Baylis found that females initiated their foraging trips based around a colony-specific bearing to their key foraging areas.

As with Australian fur seals, local conditions appear to strongly influence the foraging ranges of New Zealand fur seals. However, unlike Australian fur seals, New Zealand fur seals usually forage in pelagic waters off the continental shelf and, as a consequence, are more likely than Australian fur seals to cross the open ocean. This is evidenced by flipper-tagged individuals crossing between Australia and New Zealand and vagrants turning up on south Pacific islands, such as New Caledonia.

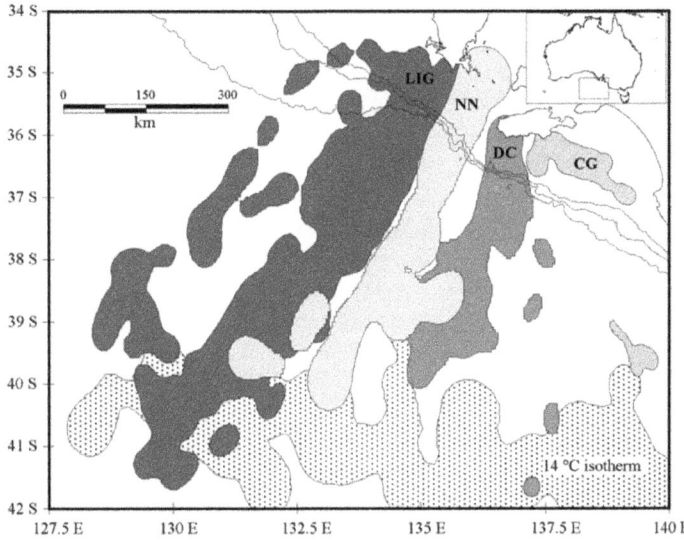

Figure 6.6. Autumn meta home ranges of female New Zealand fur seals from four breeding colonies in South Australia. (Lig = Liguanea Island, NN = North Neptune Island, DC = Cape du Couedic, CG = Cape Gantheaume). Derived from Baylis, Page and Goldsworthy (2008).

Foraging trip strategies

Trip durations by New Zealand fur seals are comparable to those of Australian fur seals, they average 7 days and range from less than 1 to more than 25 days. The durations are dependent upon how far from a rest site the seal feeds. For instance, Rob Harcourt noted that when females foraged on the shelf they undertook 3-day trips whereas, when foraging off the shelf, trips lasted over 10 days. Similarly, Brad Page and Alastair Baylis found that females foraging on the shelf had 5–7-day trips whereas those foraging in oceanic waters had 15–25-day trips (maximum 26 days). Trips by adult males to slope waters lasted approximately 9 days.

As with Australian fur seals, some New Zealand fur seals exhibit a strong fidelity to a particular foraging site. New Zealand fur seals have more choice, though, with the option to forage off the shelf if prey availability on the shelf is low. Hence, they might display less fidelity to a foraging site than do Australian fur seals. This would be difficult to measure, as fidelity is dependent on qualities of both the foraging sites, which vary enormously, and the energetic capabilities and requirements of the seals, which also vary, particularly considering the size disparity between the species.

Diving behaviour

New Zealand fur seals forage both on the shelf, where water depths range up to 200 m and off the shelf, where water depths can be over 2000 m. On the shelf, the seals may target pelagic and benthopelagic prey (as do Australian fur seals) whereas off the shelf, the seals forage in the top few hundred metres and target epipelagic prey that exhibit daily, vertical migrations. The dive strategies required to target prey in this broad range of habitats differ vastly.

To describe different dive strategies of female New Zealand fur seals, Rob Harcourt clustered dives into bouts of three or more dives. He derived three bout-types: long duration (40 min) to a medium depth (35 m), shallow (20 m) of short duration (8 min), and deep (60 m) of short duration (14 min). Bout type, bout duration and inter-bout interval all varied over time, depending on where the seals foraged and presumably what prey they targeted. Harcourt described the seal as a 'generalist' predator, as opposed to a 'specialist' predator, like Antarctic fur seals that across much of their range target predominantly one prey (Antarctic krill, *Euphausia superba*). The generalist exhibits highly flexible diving behaviours that can adapt quickly to changes in prey abundances.

Brad Page determined that the mean dive depth of adult female and male New Zealand fur seals was 41.5 m (maximum 312 m) and 52.1 m (to over 380 m), respectively. The differences probably relate to body size, with the larger males having a greater breath-hold capacity. Mean dive durations were 2.7 minutes (maximum 9.3 minutes) for adult females and 3.6 minutes (maximum 14.8 minutes) for adult males.

Diet

As could be expected for a generalist predator that forages benthically or pelagically and on or off the shelf, New Zealand fur seals prey on a broad range of species. In southern Australia, main prey on the shelf are redbait, leatherjackets, western gemfish (*Rexea solandri*), arrow squid and little penguins (*Eudyptes minor*), while the main prey in the open ocean are lanternfish (family Myctophidae) and Southern Ocean arrow squid (*Todarodes filippovae*). Other important prey types include jack mackerel, barracouta, anchovy (*Engraulis australis*), southern sea garfish (*Hyporhamphus melanochir*), swallowtail (*Centroberyx lineatus*) and calamari squid (*Sepioteuthis australis*) (Table 6.1).

Brad Page and colleagues found that the diet of adult males, adult females and juveniles all differed. They related this to the extent to which the seals foraged on or off the shelf as well as to prey size preferences. Adult males tended to consume larger prey and were more likely than juveniles or females to eat birds such as little penguins and shearwaters.

Australian sea lion

Foraging ranges

Studies utilising satellite tracking and time–depth recorders have provided a wealth of knowledge about the distance and direction of travel for Australian sea lions from many colonies (Plate 10c). They restrict their foraging to continental shelf waters, and there is marked inter-colony variability in the ranges of juveniles and adult females. The maximum foraging range of juvenile and adult female seals recorded is 118 and 190 km, respectively. Adult males range much further and have been tracked up to 340 km from their colony.

Foraging trip strategies

The foraging strategies of Australian sea lions differ from those of Australian and New Zealand fur seals in that the sea lions tend not to rest at sea. Instead, they dive continually throughout foraging trips, regardless of trip duration. Perhaps as a consequence, they spend less time at sea, around 50%, compared to 70–90% for Australian and New Zealand fur seals. Australian sea lion foraging trips are short relative to other otariids, averaging only 1.1 days (max. 5.1 days) in juveniles, 1.2 days (max. 6.2 days) in adult females and 2.5 days (maximum 6.7 days) in adult males. Generally, such short foraging trips would constrain the distance from colonies that seals can feed. To extend their foraging ranges, Australian sea lions utilise alternative haul-out locations within their foraging range to rest. Most (68%) of the sea lions tracked at Dangerous Reef and the Nuyts Archipelago used at least one additional haul-out site during the periods they were tracked. Why Australian sea lions do not rest at sea is unclear. It may be to reduce vulnerability to shark predation, because there is a high density of predatory sharks within the sea lion's range.

Nocturnal foraging appears to be common for Australian sea lions. Although adult females forage at all times of day, those at Dangerous Reef and the Nuyts Archipelago generally left on foraging trips in the early evening and returned in the morning, suggesting that night-time foraging is preferred. Simon Goldsworthy, Brad Page and colleagues tracked 30 adult females from six subpopulations within a 40 km radius in the Nuyts Archipelago. They found that females from each colony displayed one of two distinct foraging behaviours or ecotypes, targeting either inshore waters (8–20 m depth) or offshore (mid-shelf) waters (40–60 m depth). Offshore ecotype females were on average 25% heavier than inshore females (99 kg versus 79 kg), 10% longer and larger in girth. The mean heading of offshore females was SW (offshore), while for inshore females it was NE (heading to near coastal waters). Offshore females travelled 30% faster than inshore females and went twice as far on foraging tips. Mean foraging depth of offshore females was almost five times that of inshore females (50 m versus 11 m), and maximum depths obtained were almost double (72 m versus 38 m). The larger size of offshore, deeper diving females is consistent with the general relationship between body size and oxygen storage capacities observed in marine mammals.

Diving behaviour

All studies of diving behaviour in Australian sea lions indicate that they are principally benthic foragers, minimising the time spent during the descent and ascent phases of each dive in order to maximise foraging time on the seabed. Energetic studies by Nick Gales and Dan Costa suggest that Australian sea lions may be working at the upper limit of their physiological capacity while diving, implying they must work hard to exploit benthic habitats.

The diving profiles of Australian sea lions are typical for a benthic foraging pinniped. They are characterised by long and regular duration dives to consecutively similar depths, with bottom time being maximised. Average dive durations of adult females are 3.3 minutes (maximum 8.3 minutes), with bottom times of 2.0 minutes (maximum 6.3 minutes) and inter-dive surface intervals averaging 1.4 minutes. Thus, more than 60% of each dive cycle is spent near their maximum dive depth. They can undertake 10 or 11 dives per hour.

Shannon Fowler and colleagues studied the ontogeny of diving behaviour in Australian sea lions at Seal Bay (Kangaroo Island). They found that 6-month-old pups were able reach maximum dive depths averaging 30 m, and that dive depth and duration progressively increased with age (e.g. 15-month-old pups reached 68 m, 23-month-old juveniles reached 78 m, and adult females reached 103 m). By 23 months, the mean dive depth of juveniles was still only 62% of adult female dive depths. Because of the Australian sea lions' propensity for benthic diving, diving depth can be inferred from satellite tracking studies, which suggest that the maximum dive depths (based on seabed depth) where juveniles, adult females and adult males forage rarely exceed 90, 120 and 150 m, respectively.

The dive depths and foraging behaviour of Australian sea lions vary markedly between colonies and appear largely governed by the water depths available. As indicated above, where a mixture of depth ranges occurs within the foraging range of colonies, females may specialise into inshore (mean dive depths 8–20 m) or offshore (mean dive depths of 40–60 m) foraging ecotypes. In other regions, such as Spencer Gulf, adult females are restricted to shallow depths (averaging 30 m). Richard Campbell also detected very shallow mean dive depths (7–20 m) of adult females in shallow water on the west coast of Western Australia, and deeper dive depths (24–57 m) for adult females breeding off the south coast of Western Australia.

Goldsworthy and colleagues determined that the body mass of 64 adult females was significantly positively correlated with the proportion of time spent at sea, mean travel speed, mean total distance travelled, and mean and maximum foraging depth. In addition, body length was positively correlated with mean travel speed, mean total distance travelled, and mean and maximum foraging depth.

Diet

The diet of the Australian sea lion is poorly understood and based predominantly on limited qualitative and anecdotal accounts. Unlike the diets of fur seals, which can be analysed through examination of the remains of hard parts in scats, few identifiable hard parts remain in Australian sea lion scats. Feeding trials have demonstrated that most fish ear bones (otoliths) and a large percentage of cephalopod beaks are completely digested in transit through the digestive tract. The presence of stomach stones (gastroliths) is thought to be partly responsible for this. Most otariids ingest stones, but Australian sea lions take in particularly large ones (7 cm or more in diameter). They can have several of them, and they almost always retain them in their stomachs.

Because Australian sea lions range from the warm-temperate waters on the west coast of Western Australia to the cooler waters of the southern coast of South Australia, prey selection is likely to vary geographically. Most information comes from the cooler parts of their range, which is also where the sea lions are most abundant.

Rebecca McIntosh and colleagues examined the diets of Australian sea lions at Seal Bay based on digestive tracts of individuals that died (mostly drowned in fishing nets) and regurgitates (Table 6.1). The most numerically abundant prey types were cephalopods: octopus (*Octopus* sp.), giant cuttlefish (*Sepia apama*), arrow squid and calamari squid. Remains of fish species included leatherjacket, flathead (*Platycephalus* sp.), swallowtail, common bullseye (*Pempheris multiradiata*), southern school whiting (*Sillago flindersi*) and yellowtail mackerel (*Trachurus novaezelandiae*). Southern rock lobster (*Jasus edwardsii*) and swimming crab (*Ovalipes australiensis*) carapace fragments, little penguin feathers and bones, and shark egg cases (oviparous species and *Scyliorhinidae* sp.) were also identified. Most of the prey species are benthic and nocturnal, corroborating the benthic dive records and emphasis on night-time foraging.

Crittercams were deployed by Brad Page, Simon Goldsworthy and Andrew Lowther on adult female Australian sea lions in 2009–10 (Plate 11a). Prey captures recorded included octopus, rock ling (*Genypterus tigerinus*), leatherjackets, crabs and stingrays. The footage provided information not just on prey species targeted, but also on foraging habitats and strategies. It highlighted that Australian sea lions forage in a high diversity of habitats. One adult female with instrument attached at Lewis Island in southern Spencer Gulf foraged in rocky reefs adjacent to the mainland. On the first day she methodically searched under rocks and ledges to target rock ling; on the second day she switched foraging strategies to dive repeatedly to an overhung rock on the seabed, where she sat motionless. On sighting a leatherjacket, she performed short, high-speed pursuits. This 'sit and wait' predation had not been observed before in sea lions. Most other females foraged over unconsolidated benthos, which dominates the seafloor of Spencer Gulf. Sandy riffles are interspersed with swales of algae, seagrass and sponges. Here, Australian sea lion females tend to cover a lot of ground while foraging, searching for cryptic prey partially buried in sand, such as flatfish, rays and octopus. Females have also been observed pushing over flat rocks, often several times each before capturing an octopus or fish.

7

POPULATION BIOLOGY

Population monitoring: what to count?

The population size of at an otariid colony is difficult to estimate as individuals are constantly coming and going. The proportion present at a given time changes in a multitude of cycles, including long-term trends, the breeding chronology, seasons, prey availability, weather patterns, time of day, and lunar and tidal cycles. Young pups, however, are confined to the colony and the number of pups present provides a useful index of total population. Pups also are easily distinguished from other age classes and are produced during a defined breeding period.

A caution to using pup numbers to monitoring populations is that pups available to be counted fluctuate over time. Within a season, there is a high level of newborn pup mortality. At the Seal Rocks Australian fur seal colony during the 1970s, for example, Bob Warneke estimated that on average 15% of pups died in their first 2 months of life. Pup numbers present immediately after the pupping period differ from those available 1 month later. Pup numbers also change among years, even in a stable population. This is because fluctuations in prey availability influence both the proportion of females that can carry a pregnancy through to term and the survival of newborn pups. Large seas related to summer storms can cause high levels of pup mortality, particularly at low-lying colonies, in some years but not in others. Steve Kirkman, in assessing trends in pup production by Cape fur seals from a 40-year data set, concluded there had been little change in overall population despite the number of pups counted at individual colonies in consecutive years fluctuating by up to 50%. Long-term, routine counting of pups is the key to establishing an otariid's population status and trends. Whenever possible, monitoring at each colony should apply the same or comparable techniques, be conducted as soon as possible after the end of the pupping season and be repeated on about the same date across years.

The accuracy of pup counts depends on the number and movement of individuals present, the terrain and the counting method. Prior to the mid-20th century, numbers at sites in southern Australia were occasionally guessed, with figures like 'thousands' or 'at least 600' being typical. The accuracy of the guess varied depended on the ability, opportunity and motivation of the observer. Motivations could vary; for instance, the person might have wished to be as accurate as possible, or emphasise low numbers that warranted no interest from other sealers, or high numbers that required culling.

Contemporary monitoring techniques vary depending on the time available during colony visits, scientific endeavour, funding levels and disturbance levels to the seals that are considered acceptable. Three techniques are currently utilised. In order of greater precision (as well as increasing disturbance) these are aerial photography-based counts, ground-based counts (often referred to as direct counts), and capture-mark-resights. Aerial photography under-records pups: it misses those that are hidden from view, as well as those in clusters or in the water. Ground counts utilise a greater search effort but still only estimate the visible portion of the pups present. Due to the cryptic resting and mobile behaviours of the pups, an unknown proportion will always be obscured from the counter. Capture-mark-resights are the most accurate technique as they draw on all pups present and allow calculation of confidence limits around estimates, which is important when trying to determine the significance of trends in pup production across years.

A feature of pup counting has been that as more surveys are conducted and techniques are improved the apparent number of pups tends to increase. Distinguishing between increases due to improvements in techniques and population change has been difficult for some species and sites.

The short, synchronous breeding seasons of fur seals allows for the use of capture-mark-resight methods, as most pups are available for survey at the end of breeding and all are easily recognisable. However, the Australian sea lion with its very protracted breeding season presents a challenge, as some of the early born pups will have fully moulted, dispersed or died by the end of the season. Methods using multiple, sequential capture-mark-resights, spaced throughout the breeding season, can provide a means to estimate the numbers of pups present at the time of survey, and the number of births, deaths and temporary dispersals between surveys. From this, a construction of a cumulative curve allows an accurate estimate of total pup production.

Trends in numbers

Australian fur seal

Bob Warneke and Peter Shaughnessy in 1985 reviewed historic data and sealers' records to obtain an idea of Australian fur seal populations prior to the late 1700s. Based on a yield of over 300 000 skins up to 1825 and assuming about two-thirds of these were Australian fur seals, they suggested the pre-harvest population may have produced 20 000 to 50 000 pups annually, with pupping at about 23 sites. By 1825, Warneke and Shaughnessy suggested the seal numbers had been reduced by about 95%.

Seal numbers remained low for the next 150 years (Figure 7.1). The most comprehensive information throughout this period is for Seal Rocks, Victoria. This is due to its relative accessibility on the central Victorian coast and because Bob Warneke made considerable efforts to track down reports on seal numbers for this site. In December 1801, at the commencement of sealing, John Murray (from the *Lady Nelson*) visited Seal Rocks and reported 'several thousand pups on main beach'. Murray found evidence of sealing activity (remains of a camp fire and tools) that possibly occurred within

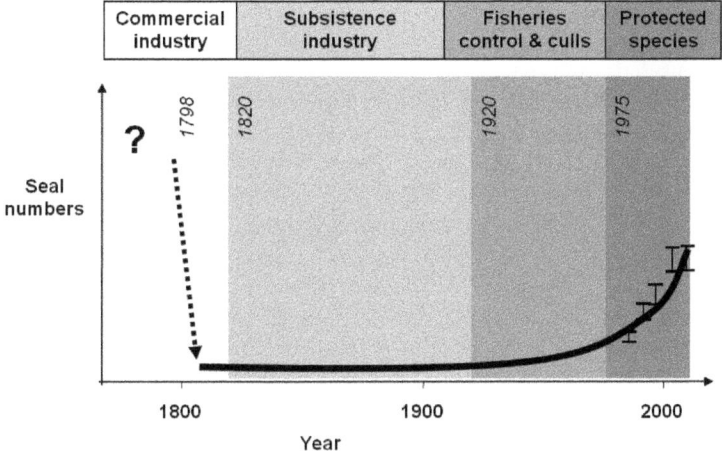

Figure 7.1. Stylised representation of the trends in numbers of Australian and New Zealand fur seals in Australian waters, in relation to species utilisation and protection.

months of his visit. In January 1813, '739 skins' were taken in a 4-day period. Sealers lived on the rocks at various times and in the 1820s the Hobart Town Gazette reported that '1000 plus skins were taken annually'.

Thereafter, reports were of 'two to three dozen seals left' in the 1850s, 'colony estimated at 1000' in the 1860s, and 'not more than 20 seals present' in 1868. In 1891, legislation was enacted in Victoria to protect the seals, and sealing ceased completely in 1923. Reports suggest that in 1903 there were '2000 to 2500', '5000 to 6000 with 600 pups' in 1928, '3000 to 5000' in 1942 and '3000' in 1945. Several seal culls were undertaken by fishermen at Seal Rocks in the early 1900s.

In April 1945, Stanley Fowler of the CSIRO over-flew much of the Australian coastline to photograph fisheries resources. In doing so, he photographed seals present at the nine extant Australian fur seal colonies. Results were not published but were retained in CSIRO archives. Bob Warneke analysed the photographs and, based on monitoring at Seal Rocks between the 1960s and 1970s, applied multipliers to estimate numbers of seals that could have been present in the breeding period and then pup numbers. Accordingly, the 11 050 or so seals present at sites in April equated with 20 000 seals in total and an annual production of 9240 pups. In April 1975, Rod Pearse of Tasmanian Parks and Wildlife repeated the aerial photography at Tasmanian sites only and found similar numbers to those present in the 1945 photographs. Based on Pearse's counts, Bob Warneke again estimated pup production for the preceding pupping period, and derived a total of 9430. This suggested there had been little population change in the intervening 30 years. In his own research at Seal Rocks between 1968 and the early 1970s, Bob Warneke estimated pup production based on marking and direct ground counts. He recorded 2100 to 2200 annually. In December 1986, Warneke conducted a species-wide survey and counted pups visible in aerial photographs. Based on work at Seal Rocks, he adjusted the counts to account for pups unseen, and estimated there was total of approximately 7950 pups present at the nine colonies. Continued annual monitoring at Seal Rocks by Warneke through to 1991 demonstrated there was considerable variation in pup numbers but, overall, there was about a 2% annual rate of increase.

Thus, from the limited data, it appears that when sealing continued to supply local markets during the 1800s, Australian fur seal numbers remained at an unknown level, but generally low. The population then increased following the suspension of sealing in the late 1800s and early 1900s, to a population that produced fewer than 10 000 pups annually, and this persisted through to the mid-1980s. Based on the pre-sealing numbers estimated by Warneke and Shaughnessy, this level could have been between half and one-fifth of the pre-exploitation level. In an assessment of population demographics during the 1970s, based on the ages of seals collected at Seal Rocks by Bob Warneke, John Arnould found an anomalous age structure. Female survivorship was low and apparently density-independent (not related to the number of animals in the population). Arnould and Warneke suggested that ongoing lethal interactions with fishing operations could account for the observed structure. Indeed, anecdotal evidence suggests that shooting of seals at colonies and resting sites occurred frequently through to at least the early 1980s.

During the late 1980s and 1990s, live pups present were estimated on a near-annual basis at colonies in Tasmanian waters (by David Pemberton and colleagues) and occasionally at Victorian colonies (initiated by Peter Shaughnessy). In addition to births at colonies, incidental pup births were recorded at about 10 haul-out sites. Considerable inter-annual variability was evident in pup numbers, especially at sites that are susceptible to storm waves, and even at more secure sites, such as Kanowna Island (Figure 7.2). Estimates of total live pups based on the counts were approximately 12 000 in the late 1980s, and 17 000 in the mid to late 1990s. In 2002 and 2007, concerted species-wide estimates of live pups, coordinated by Roger Kirkwood and colleagues, estimated totals of 21 500 and 21 900, respectively. The evident increases over time were influenced by improved coverage and censusing techniques, but undoubtedly also reflected a real growth. Thus, after remaining at half to one-fifth of the pre-harvest levels for most of last century, pup production doubled throughout a 10-year period, at a rate of approximately 6% per year, to arrive in the early 2000s at the low end of the range speculated for the pre-harvest population. Also, following confinement to nine colonies for almost 200 years, five new colonies were recognised in the first 10 years of the 21st century, including an expansion in the breeding range into South Australia.

New Zealand fur seal

In New Zealand prior to the arrival of Polynesians (Maori) in about 1000 AD, New Zealand fur seals had bred at sites around the South and North islands. Subsistence hunting progressively reduced the seals' range until colonies remained only in the south-western South Island. There at Dusky Bay in 1792 (6 years prior to harvests in Bass Strait), William Raven in the vessel *Britannia* out of Port Jackson put a gang of sealers ashore. In 10 months, the sealers collected 4500 skins. This was considered to be a poor return and further sealing ventures to New Zealand were not undertaken until 1802, which was after the initial onslaught into Bass Strait. John Ling, in his review of seal skin cargos from the region, presented totals most likely to have been New Zealand fur seals of approximately 100 000 from South Australia, 8000 from Western Australia, 330 000 from New Zealand coasts, and 487 000 from the New Zealand subantartic islands (360 000 coming from the Antipodes Islands). This excludes Macquarie Island

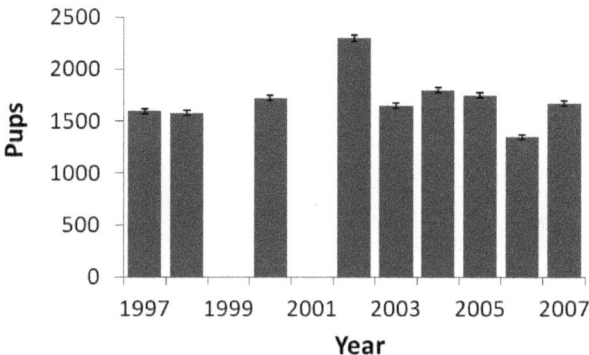

Figure 7.2. Pup production index (ground counts) at Kanowna Island, 1997 to 2007. Derived from Gibbens and Arnould (2009b). © John Wiley and Sons.

where 200 000 skins were taken of a species that may or may not have been New Zealand fur seals (they might also have been subantarctic and/or Antarctic fur seals). In addition, at least 9000 New Zealand fur seal skins came from Bass Strait (being the initial harvest at Cape Barren Island in 1798). With over 930 000 skins taken during commercial sealing, compared with the 240 000 or more Australian fur seal skins, the pre-sealing population of New Zealand fur seals was probably several times that of the Australian fur seal. In fact, Rhys Richards in 1994 estimated the pre-harvest population to be 1 250 000, based on sealers' records and including the Macquarie Island fur seals. A total population of this size would produce about 250 000–300 000 pups annually for New Zealand, Australian and subantarctic regions.

As with the Australian fur seal, recovery of New Zealand fur seal populations was slow, with most occurring since the early 1980s. Peter Shaughnessy and colleagues have closely monitored the growth in pup production at many sites in South Australia since 1988 (Figure 7.3). On Kangaroo Island between 1988 and 2010, pup production increased exponentially at approximately 11.9% per year, representing a 10.7-fold increase in 23 years. At Cape Gantheaume, pup production has increased from 457 in 1988–89 to 4632 in 2010–11, equivalent to 10.8% per annum increase. Surveys undertaken at the Neptune Islands after the 1990s indicated a lower rate of increase of 4.3% per year. Overall, the rate of increase for populations in SA during the 1990s and early 2000s averaged about 6.8% per year.

By 2011, there were at least 40 breeding colonies for New Zealand fur seals in Australia (18 in South Australia, 17 in Western Australia, four in Victoria and one in Tasmania). Most of the pup production (84%) was in South Australia, where a census estimated a total of 17 600 live pups for the 2005–06 pupping season. Most pups were born at the Neptune (48%), Kangaroo (40%) and Liguanea islands (12%). The area bounding these sites, the south coast of Kangaroo Island and south of Spencer Gulf, has the greatest density of breeding seals in Australia. Part of the explanation of this concentration is probably related to the region's proximity to highly productive shelf waters.

Numbers of New Zealand fur seals in the New Zealand region also increased after the 1970s, including recolonisation around the South Island. As the species recolonised its former range of just 1000 years ago, former sites were used as haul-outs, then developed into breeding colonies. The range expansion and population growth resulted

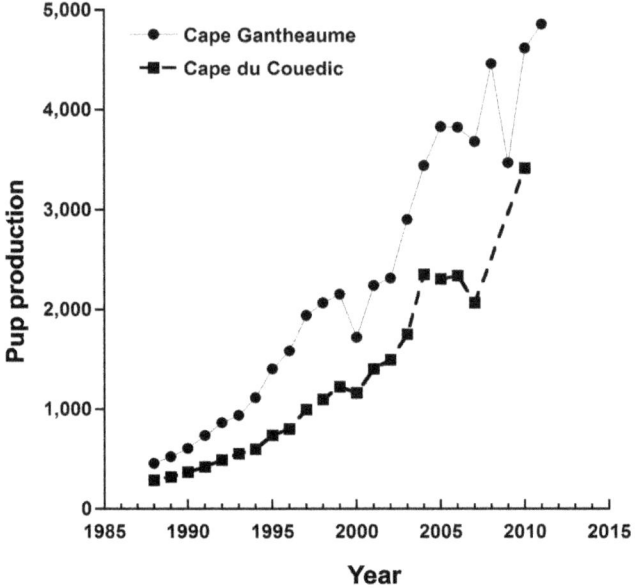

Figure 7.3. Trends in numbers of New Zealand fur seal pups at two sites on Kangaroo Island, 1988–2012. Derived from research led by Peter Shaughnessy and Simon Goldsworthy.

from both immigration and recruitment. In the New Zealand region in the early 2000s, the greatest fur seal numbers appeared to be at the Bounty Islands. There, '4380 pups and increasing' was reported by Rowland Taylor in 1980. Large colonies were also present at Antipodes, Auckland and Campbell islands. Numerous colonies dotted the coast of the South Island, the largest ones being at the Catlins in the south-east corner, Solander, Chalky and Open Bay islands on the west coast, and Ohau Point on the east coast. Occasionally, pups were also born at sites up the west coast of the North Island. At a very rough guess, as comprehensive data are unavailable, the total pup production in New Zealand waters in the early 2000s could have been similar to that for the species in Australian waters – about 20 000 per year – and possibly was increasing.

Australian sea lion
It is impossible from historic accounts to reconstruct the size and range of Australian sea lion populations prior to European colonisation. Lower prices for Australian sea lion skins and a difficulty in accessing their small, isolated populations might have meant that sealing activity had a lesser impact on them than on fur seal populations. However, they did form part of the take. For example, 200 'hair' seal skins were included with the first harvest of fur seals from south-eastern Bass Strait in 1798. It is not known if pupping occurred in Bass Strait. Certainly, sea lions frequented numerous sites around Bass Strait prior to the sealing, and were eliminated there by the sealing. They have not re-occupied this region, which had represented the eastern extent of their range.

The Australian sea lion is extremely difficult to census because many of the populations are scattered on remote offshore islands and the non-annual, asynchronous and protracted breeding season means that there are few data sets available for investigating long-term population trends. Further, because there is no other seal species similar to the Australian sea lion, limited insights into its population ecology cannot be gained from studies on other pinnipeds.

The first species-wide estimate of pup production for a breeding cycle (17.5 months) was conducted by Nick Gales and others in 1990, resulting in 2432 pups. In 2006, Peter Shaughnessy and colleagues estimated there were 4125 pups. The apparent increase was mostly due to the addition of new colonies and improvements to estimation procedures, particularly the use of capture-mark-resight procedures in large colonies and the coinciding of visits for direct counts when the maximum numbers of pups were ashore. There is little quantitative data from which to determine whether the overall population changed between the two estimating periods (or even since sealing ceased!). Potentially, numbers increased at some sites, decreased at others and some remained stable.

In the early 2000s, there were 76 confirmed pupping sites of Australian sea lions, 48 (63%) in South Australia and 28 (37%) in Western Australia. Based on the most recent estimates, 86% of pups (3622 per breeding cycle) were in South Australia and 14% (503 pups) in Western Australia. Mean pup production per Australian sea lion colony is almost an order of magnitude smaller compared to that of fur seal colonies, i.e. it is in the tens, whereas for fur seals it is in the hundreds or thousands. There were only eight colonies that are known to produce more than 100 pups per breeding cycle, all in South Australia: North Page, South Page, Seal Bay (Kangaroo Island), Dangerous Reef, Lewis Island, West Waldegrave Island, Olive Island and Purdie Island. Sixty per cent (2184) of pups were born in these eight colonies. A further 21 sites had between 25 and 100 births per cycle and the remaining 47 each produced fewer than 25 pups per cycle. Dangerous Reef produced more than twice as many pups as any other colony, and more pups than were born in all Western Australian colonies.

There is limited information on trends in abundance of Australian sea lions with the most robust time series coming from Seal Bay (Kangaroo Island). There, pup numbers decreased by about 11% over 22 years (1985–2006). In contrast, numbers of pups born at Dangerous Reef were relatively stable before 2000, but then increased markedly (it is the only colony known to have increased in that period). At North and South Page islands from the mid-1980s, through to at least 2010, pup production remained stable.

Population genetics

Genetic variation and subpopulation structure have important implications for a species' survival and evolutionary potential. Accordingly, knowledge of levels of genetic diversity and degrees of separation into distinct subpopulations aids better conservation and management outcomes. One strong consideration for the genetic status of otariid populations is that the overexploitation by sealers during the 1800s probably pushed many populations through a 'genetic bottleneck'. This reduces the genetic diversity and alters the subpopulation structure that was present prior to the exploitation.

Researchers can trace genetic lineages by examining nuclear DNA microsatellites, inherited from both parents, and mitochondrial DNA (mtDNA), which is inherited maternally. A combined analysis of both DNA sources allows an interpretation of paternal lineages. In mammals, females tend to display a greater degree of natal site fidelity than do males. This behavioural distinction can result in strong population structure at the mtDNA level but little or no structure at the microsatellite level.

Australian fur seal

Australian fur seals originated from a trans-Indian Ocean migration by Cape fur seals from southern Africa, less than 18 000 years ago. Due to this relatively recent founding, the genetic diversity and subpopulation structure of the Australian fur seal are limited. The overexploitation by sealers probably resulted in a second bottleneck of the genetic diversity of this otariid.

Several studies have compared genetic material from Australian fur seals with that of other otariids in examinations of evolutionary trends. Only a study by Melanie Lancaster and colleagues has examined the genetic structure within the Australian fur seal (Figure 7.4). Using microsatellite loci (regions of short repeat sequences that can vary in number of repeats) on nuclear DNA and mtDNA analysis of skin tissue collected from pups at nine separate colonies between 1992 and 2002, Lancaster revealed no between-colony differences in allelic (one of two or more versions of a gene) diversity or microsatellite heterozygosity (five loci) and no differences in haplotype (a

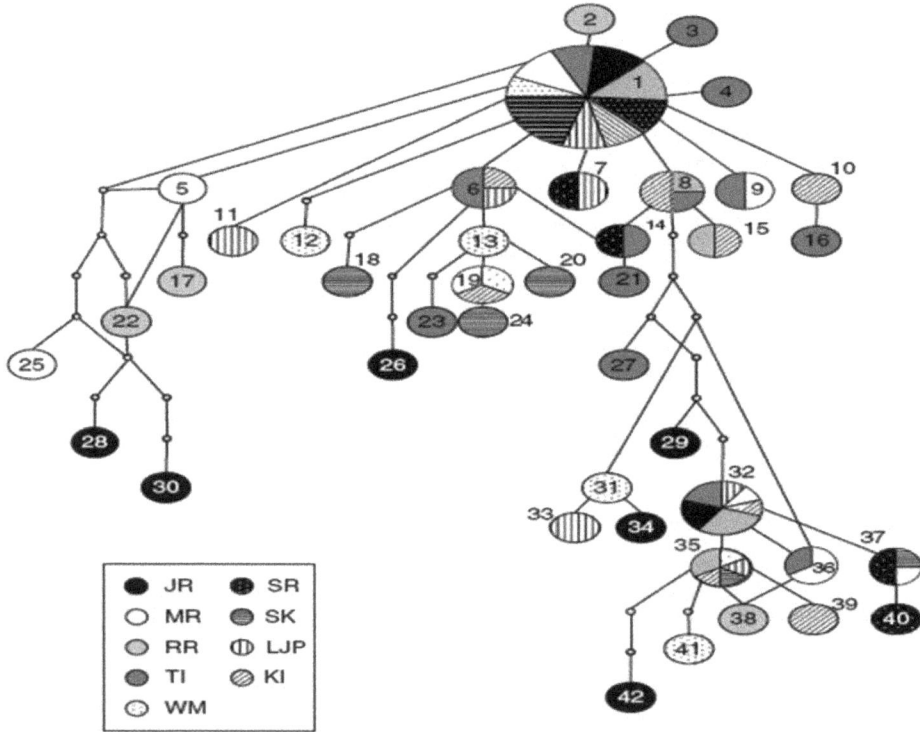

Figure 7.4. Australian fur seal mitochondrial DNA haplotype network, where numbers designate unique haplotypes and each single line represents one mutation between haplotypes. Small open circles indicate missing haplotypes. The size of each circle reflects the number of individuals sharing a haplotype; patterns represent different Australian fur seal colonies. Derived from Lancaster, Arnould and Kirkwood (2010). © John Wiley and Sons.

Table 7.1. Comparison of genetic diversity within the Australian fur seal and other fur seal species, indicating its relatively low levels, potentially reflecting recent establishment. Derived from Lancaster, Arnould and Kirkwood (2010). © John Wiley and Sons.

	Alleles per locus		Heterozygosity (H_0)	
	Mean	Standard deviation	Mean	Standard deviation
Australian fur seal	8.0	3.1	0.58	0.26
Antarctic fur seal	15.0	7.9	0.71	0.17
Subantarctic fur seal	17.2	7.5	0.82	0.08
New Zealand fur seal	16.8	5.9	0.82	0.06
South American fur seal	9.0	4.6	.	.

group of genes which are inherited together) diversity within the mtDNA control region (Table 7.1). Although there was some evidence for isolation between colonies by distance, analyses supported a single population. Gene flow among colonies appeared to be substantial and facilitated by both sexes, indicating that the Australian fur seal population could be considered a single, genetically panmictic unit (one with no mating restrictions).

High mitochondrial haplotype diversity but low nucleotide diversity indicate that Australian fur seals have experienced a recent founding followed by rapid expansion. Similar signatures of expansion have been observed in the mtDNA control region of other otariids including Cape, Antarctic, subantarctic and New Zealand fur seals, suggesting the patterns are related to the overharvesting that all these species experienced around 200 years ago. However, unlike Australian fur seals, several species display a degree of mtDNA structuring, indicating subpopulation distinctions. The relative degrees of population structure in otariids may be due to the extent of geographic range, with the range of Australian fur seals being small relative to other fur seals.

New Zealand fur seal

In a study of blood serum protein variation in southern fur seals published in 1970, Peter Shaughnessy compared transferrin type proteins in sera from New Zealand fur seals in Western Australia and New Zealand. All proteins found in Western Australia were also present in New Zealand samples, but not all New Zealand transferrin types were in the Western Australian samples. This suggested that the New Zealand fur seals in Western Australia had arrived from New Zealand, rather than the reverse. Later, Gina Lento and co-workers compared New Zealand fur seal populations using mtDNA and found significant differences between those in Western Australia and those in New Zealand, suggesting that the colonisation of Western Australia had not been a recent event – it was a recolonisation from local stocks rather than from New Zealand following the near extirpation by sealers. Lento found that New Zealand fur seals in Tasmania and Bass Strait possessed haplotypes from both extremities of their range.

New Zealand fur seals display similar levels of genetic diversity in their microsatellite loci to other otariid species. Bruce Robertson and Neil Gemmell suggest that this, and other genetic features, such as considerable and non-homogenous allelic diversity, moderate levels of inbreeding, and consistent departure from Hardy–Weinberg proportions for all loci, all point to strong population substructuring in the fur seals. Contrary to this expectation, however, breeding colonies around the New Zealand South Island were generally weakly differentiated based on microsatellite structure, possibly due to consistent gene flow between them. Differentiation was strongest on the west coast, where 41% of individuals could be assigned to their colony-of-origin and almost 70% of individuals to their region. On the east coast, a population at Horseshoe Bay, on Banks Peninsula, appeared distinct from others, suggesting different origins. Potentially, most colonies were recolonised by local 'spillover' from other mainland sites, but the Horseshoe Bay colony was colonised by a subantarctic clade, which was in the process of being obscured by gene flow from neighbouring colonies.

Australian sea lion

Richard Campbell and co-workers utilised mtDNA and nuclear (microsatellite) DNA markers to investigate population structure of the Australian sea lion. Samples collected from eight colonies in Western Australia and two in South Australia provided evidence for strong sex-biased dispersal, manifested primarily in extreme female natal site fidelity (or philopatry), which is unparalleled among otariids. Population subdivision was evident at both large and small geographic scales, with some fixed differences between breeding colonies separated by even short (20 km) distances. They also detected high levels of fixation among mtDNA markers in many of the small Western Australian colonies, which was attributed to high rates of genetic drift. The extreme female natal-site fidelity and a high risk of extinction of smaller colonies from stochastic processes have significant conservation and management implications, including the need for a colony-specific management approach.

Recently, Andrew Lowther and colleagues undertook additional investigations into the genetic population structure of South Australian colonies (Plate 11b). They sampled mtDNA from 17 colonies and identified 21 unique markers (haplotypes), of which about half were unique to a particular colony. They found strong genetic partitioning among colonies with most being characterised as individual populations. However, The Pages (the North and South Page islands), which are under 2 km apart, shared mtDNA haplotypes in similar proportions so were considered one colony. Also, Dangerous Reef and English Island in Spencer Gulf grouped together, as did six colonies in the Nuyts Archipelago. Results corroborated the findings of Campbell, confirming a strong genetic subpopulation structure within the South Australian population of Australian sea lions.

In contrast to the mtDNA results, a dispersal pattern was evident in the microsatellite markers, which are paternally derived, suggesting there was a high degree of male dispersal. The levels of male dispersal appeared to be sufficient to overcome female philopatry, making small groups of colonies effectively panmictic. However, the range of male dispersal appeared to be limited to approximately 200 km. This results in regional population subdivisions reflecting geographic distance, as well as the temporal processes of extinction and colonisation.

Population sizes

In addition to monitoring status and trends, it can be valuable in wildlife management to have an estimate of the total population size of a species. Scientists are reluctant to provide such estimates, and couch them in broad ranges or with a series of provisos. This is because extending live pup numbers to provide population estimates involves a range of assumptions. For example, multiplication factors for biases inherent in counting techniques, proportions of pups born that were not available to be counted, pregnancy and pupping rates, survival information for each sex, inter-colony and inter-annual variability, and assumptions about whether populations are increasing, decreasing or stable. The estimates can be subjected to repeated modification as new data are collected.

In 1964, British researcher Humphrey Hewer published life tables for grey seals (*Halichoerus grypus*) based on a sample of 80 females and estimated that the total population size was approximately 3.5 times annual pup production. In 1978, John Harwood and John Prime assessed survival and fecundity in a sample of over 1000 male and female grey seals and refined the multiplication factor. They proposed that population sizes of most polygynous pinnipeds were likely to fall within the range of 3.5 to 4.5 times pup production. Since its derivation, this range has been broadly applied to many pinniped populations. The relationship between pup production and population size varies between species, however, and is dependent on species-specific demographic parameters (such as survival and pupping rates) and population trends, which vary over time. Species-specific multipliers have since been estimated for several otariids and fortunately these include the Australian species: 4.5 for Australian fur seals, 4.23 and 4.9 for New Zealand fur seals and 3.8 to 4.8 for Australian sea lions.

A further caution to obtaining population estimates from direct counts of live pups is that such counts will underestimate total pup production. An unknown and variable proportion of pups die, disintegrate and are washed away during and shortly after birth and are unavailable to count. To record total births requires a near-daily monitoring, which is not possible at most colonies. Furthermore, pup mortality rates will be colony-specific, change through the pupping period and vary between years, so cannot be accurately transposed to other sites and years. Having stated this, an estimate of pre-count pup mortality can improve the estimate of pup production. In Australian fur seals, the figure of 15% pup mortality prior to counts has been applied. This figure represented the mean mortality to early January during Bob Warneke's multi-year study at Seal Rocks in the 1970s and, perhaps coincidently, was also the figure recorded in 1992 at Tenth Island by David Pemberton.

Multiplying the 2007 estimates of live pup numbers at each site of Australian fur seals by 1.18 to account for pup mortality and hence estimate pup production, and by 4.5 to estimate total population size (including pups), provides a rough estimate of 110 000 to 120 000 Australian fur seals. Estimates from the mid to late 2000s of New Zealand fur seal pup production in South Australia (17 600), Western Australia (3100) and Victoria and Tasmania (250) suggest an Australia-wide pup production of around 21 000. Similar numbers were thought to exist in New Zealand waters. Multiplying a total of 40 000 by 1.18, to include unobserved pup mortalities (recognising that this multiplier would be inaccurate for this species), and by species-specific multipliers of 4.23 and 4.9, suggests a total population of 200 000 to 220 000 individuals. Based on the estimate of 3622 Australian sea lion pups born in 2006, Peter Shaughnessy and colleagues recently estimated the species's total population was approximately 14 780 individuals.

In the early 2000s, Australian and possibly New Zealand fur seals in Australia could have been approaching the lower end of the ranges of numbers estimated to be present prior to sealing. In contrast, New Zealand fur seals in the New Zealand region were still less than a quarter of the estimated, pre-harvest, population levels. Both fur seal species appeared to be increasing in numbers and support panmictic populations, although there were varying degrees of genetic separation based on geographic distance.

Prior to sealing, Australian sea lions were present in their hundreds in Bass Strait. By the early 2000s, there had been no reoccupation of Bass Strait. The lack of return to former sites, a strong subpopulation structure, small numbers and low rates of fecundity highlight the species's high conservation risk. This has lead to it being listed as Threatened (vulnerable category) under the Australian Government's *Environment Protection and Biodiversity Conservation Act 1999*, and as Endangered under the International Union for the Conservation of Nature's (IUCN) Redlist.

Population regulation

Population sizes are regulated by density independent (such as climate) and dependent factors (such as competition for breeding space and food). The factors interact and are influenced by genetic adaptation. In combination, they result in an amalgamation of inter-annual, multi-decadal and longer term trends and cycles of abundance. Add human-induced population regulation and it becomes very difficult to recognise degrees of stability.

Seals were subjected to uncontrolled over-harvesting from southern Australian waters by humans in the early 1800s and then held at very low levels through continued shooting, mostly by fishermen, until the mid to late 1900s. After the 1970s, the fur seal populations increased, in part as a response to receiving legal protection in 1975. Future levels attainable by the populations will depend on ongoing regulating processes, including changes to marine food webs by commercial fisheries and climatic changes. Here, we detail two other regulating mechanisms – diseases and parasites – which may influence future population sizes of seals in southern Australia. Both relate to interactions with other species, and may be regulated by density dependent and independent mechanisms.

Diseases and pathology

Like most mammals, seals are susceptible to a wide range of pathogens, including bacterial, viral, fungal and parasitic infections (see the next section on parasites). In addition, traumatic injuries are a common occurrence in this group. Minor wounds usually heal quickly in seals but serious wounds or those that are located in bone or joints may become severely infected, leading to the deaths of individuals. Several diseases may be rapidly passed among individuals, and potentially lead to mass mortalities. The occurrence and severity of infectious disease outbreaks are influenced by factors that increase the likelihood of disease introduction to, and spread within, susceptible populations. These factors include aggregation of animals in colonies, especially during breeding seasons, and fluctuations in climatic and other environmental factors that influence animal foraging ranges.

Examples of pathogens that may affect seals at the population level include morbilliviruses; influenza viruses; bacterial infections such as leptospirosis, tuberculosis and brucellosis; and infections with the protozoan *Toxoplasma gondii*. Researchers may locate the actual pathogens or detect antibodies to them, which form in individuals in response to exposure. Such antibodies may remain with them for life; hence, it may not be known when the exposure occurred. One morbillivirus, known as phocine distemper virus, has resulted in mass mortality events involving many thousands of phocid seals (mostly grey seals) in the North Sea. At least one of the outbreaks was believed to have resulted from seals being infected by canine distemper from domestic dogs. Mass mortalities resulting from a morbillivirus infection have so far only been seen in phocid seals. In New Zealand, however, wildlife veterinarian Padraig Duignan and co-workers reported phocine distemper-like antibodies in both New Zealand fur seals and New Zealand sea lions. Also, distemper-like virus antibodies have been reported in South American otariids. A 3-year survey for pathogens and antibodies amongst Australian fur seals, commenced by veterinarian Michael Lynch from Melbourne Zoo in 2007, found no evidence for exposure to morbilliviruses.

Influenza A viruses can cause disease in many birds and mammals and are well known for their ability to cross species barriers. One variant was the cause of a mass mortality event in harbour seals off the New England coast of North America. Low levels of antibodies have been recorded in other seals, including one otariid (the South American fur seal). They were not recorded in Australian fur seals during Michael Lynch's serological survey.

A bacterium, *Leptospira interrogans pomona*, has long been identified as a significant cause of morbidity and mortality of Californian sea lions in California and Mexico. Its prevalence fluctuates between years. The disease caused by this organism, leptospirosis, is also significant in humans, rats, cattle and dogs, and over 23 serovars (variations) are endemic in Australia. Disease transfer requires direct contact with a carrier or its urine. No antibodies to *Leptospirosis* have been recorded in Australian pinnipeds.

Bacteria from the genus *Brucella* have a worldwide distribution and marine strains have been isolated, including a seal-specific pathogen, *B. pinnipedialis*. Brucellosis in terrestrial wildlife and domestic animals causes abortion and infertility. In pinnipeds, the pathological significance of *Brucella* infection is unclear; antibodies can be isolated from healthy individuals or be associated with bronchopneumonia, abscesses and lymphadenitis in unhealthy individuals. Natural infections of marine strains of *Brucella* in humans have been reported in New Zealand and South America. Michael Lynch found a high and fluctuating prevalence of antibodies to a *Brucella*-like organism in Australian fur seals, but did not link the antibodies with any pathogenicity (Figure 7.5).

Another significant pathogen of seals is *Mycobacterium*, the genus of bacteria involved in tuberculosis infection. This is a zoonotic disease transferred through air. Tuberculosis caused by *M. pinnipedii* has been identified in Australian sea lions, Australian fur seals, and New Zealand fur seals in both New Zealand and Australia. In Australia, an outbreak of *M. pinnipedii* infection occurred in a small, captive seal population in Western Australia in the 1980s; all seals in the facility were destroyed. One handler contracted the disease. Then in 1989, an adult male Australian fur seal presented with laboured breathing and deteriorating condition at the Hobart docks, Tasmania. Upon euthanasia and necropsy, the case veterinarian, Rupert Woods, found a *Mycobacterium* infection had destroyed all but 15% of one lung and the pleural cavity was filled with fluid. In Michael Lynch's serological survey of Australian fur seals, no antibodies to pathogenic *Mycobacterium* were found and it was suspected that the disease was at a low prevalence in this species.

Toxoplasma gondii and related protozoa are occasional pathogens of marine mammals and some seal populations have a high prevalence of antibodies to this

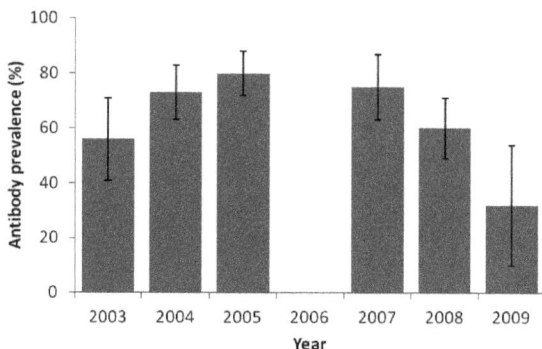

Figure 7.5. Temporal variation in the prevalence of adult Australian fur seal females on Kanowna Island that tested positive to *Brucella* antibodies. Derived from Lynch, Kirkwood *et al.* (2011).

parasite. The life cycle of *T. gondii* requires a felid (cat) definitive host with other mammals being intermediate hosts. It is intriguing that some of the highest prevalences amongst pinnipeds are in Arctic phocids, which would have minimal exposure to felids. *T. gondii* has caused meningitis in an adult Australian sea lion and severe disease in a pup of the same species, but antibodies to the disease were not isolated in the 2007–09 serological survey of Australian fur seals. The prevalence of *Toxoplasma* species in the marine environment of southern Australia appears to require further investigation.

Since 1989, a potentially serious alopecia (hair loss) syndrome has been recognised in Australian fur seals. The condition manifests as bilaterally symmetrical hair loss of the back and head. It was investigated by Michael Lynch and colleagues in 2009–11 (Plate 12). Prevalence was highest at the large Lady Julia Percy Island colony, off western Victoria, and rare at other sites. Juveniles were affected most frequently (up to 25%) and almost all (93%) of the juveniles affected were female. That meant that up to 50% of juvenile females at the colony were affected. A small proportion of adult females also exhibited the condition, but no adult male cases were noted. Alopecic seals were in significantly poorer body condition than non-affected animals. Due to its high prevalence and potentially high energetic costs to individuals, this syndrome could regulate the numbers of Australian fur seals at the Lady Julia Percy colony. Microscopic analysis by Lynch revealed that the hair was fracturing rather than being lost from the follicle, and trace metal analysis indicated that affected seals had significantly less zinc in their hair than did unaffected seals. Zinc is an important component of a protein that forms the outer cuticle of hair strands. Alopecic seals also had higher levels of toxic heavy metals (including mercury and lead) and persistent organic pollutants in their body tissues, suggesting they forage in ecosystems where concentrations of pollutants are high. Possible causes of the alopecia are being investigated.

Parasites

Otariids, like most animals, constantly host parasites, including internal worms (cestodes, trematodes, nematodes and acanthocephalans) and external arthropods (lice and mites). The abundance and impact of the parasites changes with age, season, body condition and diet of the individual. Unwell seals may exhibit large parasite loads; however, it can be difficult to determine if this is a cause or a result of the seals' poor condition. Particular species of parasites may be host-specific while others infect several species. Life cycles of the parasites typically include intermediate hosts and/or free-living phases. For example, many internal parasites cycle through the seals' prey while external parasites may live part of their life cycle in the substrate at colonies.

Cestodes (tapeworms)

Over 14 genera and 50 species of cestodes have been recorded in seals, most of which also infect other primary hosts, including other marine mammals, sharks and seabirds. Few genera are specific to seals (these include *Anophryocephalus* and *Pyramicocephalus*) and few species are unique to one seal species. The cestode life cycle involves a larval stage that usually forms cysts in the muscle of the seals' prey and transfers to the seal during prey consumption.

Adult cestodes reside in the gut of their host and are common in most seals. Mass infestations may occur in the stomach and intestine. The impact of adult cestodes on their host usually is minor. Seals act as intermediate hosts for some cestode species and in these situations tissue cysts formed following infestation occasionally affect the health of the individual. For example, a Cape fur seal once observed to be convulsing on a beach in South Africa was found on necropsy to have cysts in its brain tissue of the cestode *Taenia solium*, a human parasite that normally encysts in the muscle tissue of pigs.

Trematodes (flukes)

Trematodes inhabit visceral organs, including the intestine, liver, gall bladder and pancreas, of their host. They may occur in large numbers in individuals of some pinniped species, causing inflammation and leading to death of the host. Similarly to cestodes, approximately 19 genera and 24 species of trematodes have been recognised in seals, with few genera being unique to them (including *Pricetrema* and *Zalophotrema*).

Nematodes (roundworms, hookworms, etc.)

Nematodes are the most ubiquitous and detrimental of the internal parasites of seals. Several species are common in the stomach and gut, where they may ingest stomach contents or host tissue, while others can infest the lung (lungworm), the heart (heartworm) and other tissues. At least 15 genera and 50 species are known in seals, including three genera of roundworm (*Anisakis*, *Contracaecum* and *Terranova*) that inhabit the stomachs of all seal species. Roundworms can cause inflammation and ulceration of the stomach wall, which reduces animal health. Occasionally, roundworms can perforate the stomach, leading to peritonitis and death.

Hookworms attach to the small intestine of their primary hosts and ingest their hosts' blood. Pups of many otariids are primary hosts of several *Uncinaria* hookworms. Severe hookworm infestations can cause anaemia, enteritis, morbility and mortality. In 2002–03 at San Miguel Island, USA, a hookworm enteritis-bacteraemia complex was recognised as the main cause of mortality of California sea lion pups. Hookworms also appear to exacerbate pup mortality rates of northern fur seals, New Zealand sea lions, South American sea lions and Steller sea lions, and have been identified in pups of Australian fur seals and Australian sea lions (Figure 7.6).

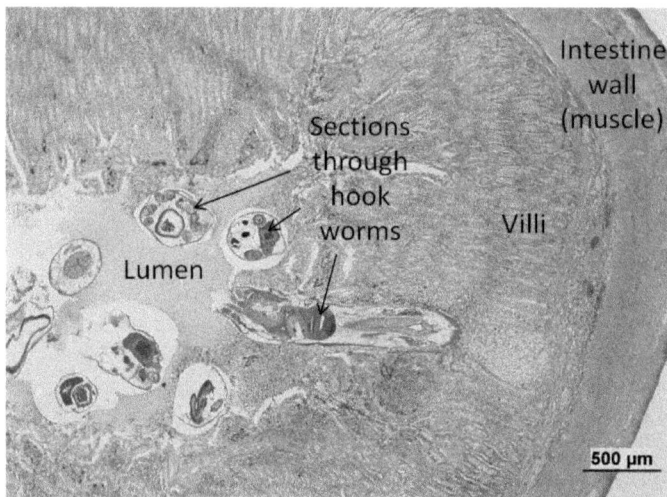

Figure 7.6. Hookworm invading through the mucosa of the small intestine of an Australian sea lion with associated haemorrhage and inflammation. Light microscope photograph by Rachael Gray.

Hookworms have a free-living larval phase which requires a soil substrate. Eggs are passed in the hosts' faeces within 2 or 3 weeks of infection and hatch after 23 days. Free-living larvae can remain in the soil for an extended period, before opportunistically burrowing through the skin of seals or being ingested. In this 'secondary' host, the larvae migrate to fatty tissue and remain in arrested development. If the seal is an adult female, the larvae migrate to the mammary glands, potentially under a hormonal signal, and pass to newborn pups via the colostrum (first milk). Degrees of hookworm infestation can be influenced by the surface structure at colony sites, as well as the density of seals.

Adult lungworms infest the lungs, bronchi and bronchioles of seals where they may cause inflammation (bronchopneumonia). Lungworm-induced pneumonia is a common contributing cause of mortality of otariids. Animals are particularly susceptible when lungworms first infect them; this coincides with weaning. Heartworms inhabit the heart and major blood vessels of their host. Several genera occur in otariids, but their biology and pathology have not been studied in depth.

Acanthocephalans (thorny-headed worms)
Acanthocephalans attach to the wall of the small intestine of their host. The larvae pass through intermediate hosts such as fish and amphipods. Two genera (*Corynosoma* and *Bolbosoma*) and over 20 species have been identified in seals. Their effect on seals appears to be minor.

Anoplura (sucking lice)
Lice are ectoparasites that may live in scurf (skin flakes) on the skin surface and feed on dead tissue, pierce and feed on live skin, or suck the blood of their host. Four genera and up to 20 species parasitise seals, and most species are host-specific. Some species appear to have evolved with their host such that genetic research into the lice can

inform on evolutionary processes within the seals. For example, the lice *Antarctophthirus microchir* infests all species of sea lion (and no other pinnipeds), suggesting it lived on ancestral sea lions prior to their groups' radiation. Rebecca McIntosh and Durno Murray found that 50% of Australian sea lion pups at some colonies were infested with lice. Pups with high infestations were similar in size and condition to those with no lice, implying the lice do not cause much trouble.

Acarina and Demodicidae (mites)
In the family Acarina, one genus of mites (*Halarachne*) is an endoparasite of phocids and another (*Orthohalarachne*) is an endoparasite of otariids. Seal mites occur in the nasal passages of their host. They may transfer between seals via nose-to-nose contact. Adult mites burrow into the nasal mucosa and feed on lymph. They may cause inflammation and irritation, and heavy infestations may affect respiration, thereby predisposing the animal to secondary pulmonary infections.

The Demodicidae are represented by the species *Demodex zalophi*, which has been recorded mainly in the hair follicles of the flippers of Californian sea lions, but has been noted to occasionally cause patchy hair loss in captive individuals.

8

CONSERVATION AND MANAGEMENT

To 'conserve' is to protect and maintain while to 'manage' is to assess, develop a strategy and do something, even if the 'something' is to decide to do nothing! Conservation and management of a species rely on the recognition of influencing factors, and determining whether and how these can be best managed or mitigated. The priority for managers is usually to mitigate anthropogenic factors.

At present, otariids in southern Australian waters generally are considered to be an important component of marine ecosystems in which humans also utilise a range of marine natural resources (e.g. fisheries, oil and gas, ecotourism, wave energy and desalination plants). As key predators, healthy seal populations to some extent demonstrate healthy ecosystems. Seals are afforded a degree of protection that is less than that of humans and cetaceans, but above that of fish and invertebrates. This has not always been Australian society's attitude toward seals, and may not be the attitude held by them in the future.

Attitudes

Prior to European settlement, Aboriginal Australians viewed all wildlife with respect and utilised it as a source of commodities, principally food. Early European settlers also valued the commodities but quickly realised the economic potential of large numbers of seal pelts and blubber. In diaries, some Europeans described their affection for the individual seals, pitied them perhaps, and recognised that the harvest was unsustainable. Generally, though, the attitude of sealers and merchants was to harvest as many seals as possible before a competitor discovered them. Sealing was a very important industry in the establishment of European settlement in Australia. Most of the commodities required for society to function at that time had to be sourced from Europe. Sealing provided the first large-scale industry exporting raw materials of high value (seal skins and oil). The product required limited processing and could be traded with South-east Asia in return for other commodities needed by the developing colony.

Throughout the 1800s and early 1900s, society continued to view the seals as a potentially harvestable resource for the future. Attempts in the late 1800s to protect the seals largely aimed at recovering a resource so it could be harvested in the future. As fin-fish fisheries developed, seal management became a balance between allowing seal stocks to recover and reducing their perceived competition with fisheries. Several government-sanctioned culls were conducted and seals were routinely shot by fishermen.

Following World War II, the development of textile and petroleum industries provided alternative and more easily obtained products than could come from seals, and interest in seal harvesting in Australia waned. Also, the increasingly educated and affluent society nursed an attitude of conservation. Wildlife warranted protection for reasons other than human use. Seals, being charismatic and playful large mammals, were appealing to human sentiments, and became popular zoo and circus exhibits. This aided a change in societal attitudes from seals being seen as a resource or competitor to them being perceived as individuals and semi-domesticated. Central (largely urban) society developed affection for individual animals and an interest in preservation of natural environments. However, the utilitarian attitude of sections of society remained. Inclusion of seals in wildlife protection Acts during the 1970s was initiated by a view that they required protection from indiscriminate shooting by fishermen.

North American researchers David Lavigne, Victor Scheffer and Stephen Kellert reviewed temporal changes in human attitudes toward wildlife and contrasted attitudes of different societal groups. Generally, the more affluent and educated the society, and the less it relied on direct economic benefits from wildlife use, the more moralistic was its attitude. Sections of a society that tended to be more moralistic were females, those with higher levels of education and urban dwellers. Scientific research and the media have strong roles in informing society and influencing attitudes toward wildlife, including otariids. They can discover, highlight and direct attention towards conservation and management issues.

Harvests and culls

In Australia, seal harvests (for a product) were conducted in an opportunistic and unregulated fashion from 1798 to the 1880s. Harvesting then became regulated under various state governments and eventually ceased in 1923. Thereafter, several government-sanctioned culls (to reduce a perceived threat) were performed by fishermen, the most recent being in Victoria in the 1940s. Incidental shooting of seals by aggrieved fishermen or for 'sport' has continued through to the present (Figure 8.1). In September 2006, for example, seal researchers working at Kanowna Island beside Wilson's Promontory witnessed fishermen from a professional fishing vessel fire over 100 rounds at Australian and New Zealand fur seals. The fishermen were charged under firearms and wildlife offenses. Generally, though, the rate of killing of seals in Australian waters has declined over time. Elsewhere, however, harvests and culls of seals have continued into the 21st century. Here we present two examples of these: takes of harp and harbour seals in the Arctic, and Cape fur seals in Namibia.

Figure 8.1. An Australian fur seal shot in the shoulder and washed ashore in Western Port, Victoria, January 2008. Photographer: Neville Johnson.

Harp seal (*Pagophilus groenlandicus*) harvesting developed from traditional hunting into a commercial, offshore industry in 1794 (Figure 8.2). Through evolutions of techniques over the following 200 years, an annual harvest of over 250 000 seals was sustained. Reasons for the sustainability of the industry included there being millions of seals in the population, their broad range, the short duration of their availability to commercial harvesting (several weeks per year), a difficulty with accessing the seals on

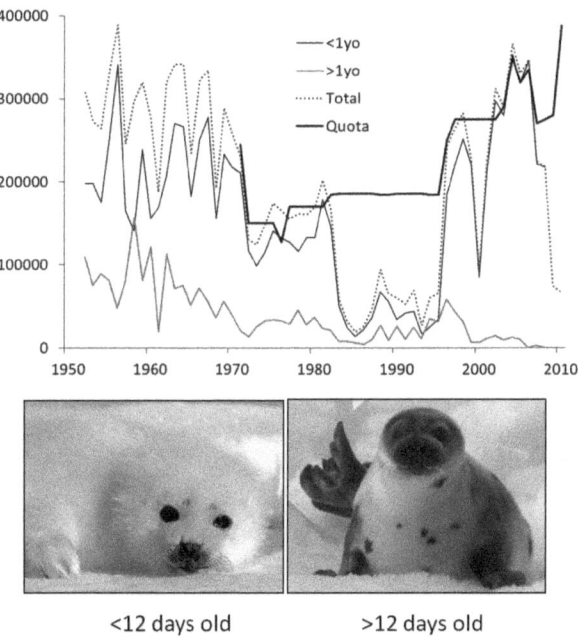

Figure 8.2. Harvests of seals in the Canadian harp seal fishery. Data are derived from http://www. harpseals.org/about_the_hunt/quota_tac.php and are reproduced with permission from Harpseals.org.

the Arctic pack ice, a continual market for the skins, and societal, financial and ethical support. During the 1970s, '80s and early '90s, harvests declined to approximately 150 000 per year, then to below 50 000 per year, due to the introduction of restrictive total allowable catches, then due to a European Economic Community (EEC) ban on the importation of seal products. The ban followed a public awareness campaign about the plight of harp seal pups. The pups initially possess a white coat, but moult after about 12 days into a mottled silver pelage. When it was realised that the EEC ban would be lifted if the seals were harvested after moulting their white coats, the total allowable catches were increased and the industry immediately returned to harvests of 250 000 or more skins per year. In 2009, the ban was reinstated, resulting in both an immediate decline in harvest and an increase in the government-set quota, to 388 000 in 2010.

After the 1980s, pro- and anti-harp seal harvest advocates argued economic, social and scientific benefits for their cases. For example, the seals were implicated by pro-harvesters as the cause for the 1992 collapse off the Canadian east coast of northern cod (*Gadus morhua*) stocks. It was speculated that the seals both ate all the cod and were a critical intermediate host to a parasite that decimated the cod. Subsequent ecological modelling exonerated the seals and implicated overfishing as the cause for stock collapses. Pro-harvesters then accused the seals of 'retarding cod recovery'. Ultimately, political expedience drove the continued seal harvesting in Canada. The seal harvests were seen to provide economic and social benefits to a section of the society that would otherwise undergo hardship, and more votes would be secured by supporting harvests than by banning them.

Harvest of seals in southern Africa has run a similar course. Cape fur seals were subjected to uncontrolled commercial harvesting from as early as 1610 to 1893, then government-controlled harvesting through to 1990 in South Africa and at least to 2012 in Namibia. Since the 1960s, the most valuable commodities of the harvests have been the cured skins of yearlings (for the fur trade) and the dried genitalia of adult males (considered an aphrodisiac in some Asian markets). Following a collapse of fur markets in the 1980s, which was caused by pressure from animal welfare and conservation groups, the industry relied on government financial support. This was sustained due to fishing industry pressure, which was fuelled by perceptions that seals threatened commercial hake (*Merluccius capensis* and *M. paradoxus*) catches, despite scientific data indicating otherwise. Subsequent ecological modelling demonstrated that the large scale culling of seals was unlikely to provide a benefit to fisheries catches. Namibia received its independence from South Africa in 1990. In that year, taking into consideration fisheries interests, scientific data, economic value and politics, Namibia continued its seal harvests, while South Africa postponed its own indefinitely.

Fishery interactions

In Australia, some of the greatest conservation and management challenges for seals relate to interactions with fisheries. These can be direct (operational) or indirect (ecosystem). Direct interactions are documented frequently because they can be physically observed and, in some instances, directly managed. In contrast, indirect interactions are largely intangible and difficult to quantify. These occur when the same prey or prey from the same food web is targeted by both seals and fisheries.

Direct fishery interactions

Seals and fishermen are active hunters of fish, cephalopods and crustaceans. Because seals may not detect the presence of fisheries equipment or because they learn that fishing operations make good foraging sites, they often interact directly with fisheries operations. These operational interactions include feeding on species surrounding or in fishing gear, feeding on bait or species that are snared in the gear, deterring target species from approaching gear, and getting trapped in the gear. Operational interactions are a nuisance that can result in financial costs to fisheries, and may cause injury or death to seals.

It can be difficult to quantify the impacts of direct seal–fishery interactions. Fishermen may perceive that documenting such interactions could reflect negatively on themselves and their industry. As a consequence, interactions such as bycatch incidents probably are under-recorded. Conversely, a fisherman who wishes to acquire government support or recognition of financial costs of seal interactions might exaggerate the amount of loss. The 'catch 22' of whether to report or not can be frustrating. Moreover, it is actually nearly impossible for fishermen to prevent the seals from interacting with their gear or to accurately quantify how much the seals take from their fishing gear. It is also extremely difficult to quantify the numbers of seals involved and the numbers that may be injured or killed (e.g. entangled, or drowned during interactions). Fishing logbook data and anecdotal accounts can signal the potential for interactions. Accurate quantification of interactions can come only from unbiased, experienced fishery-independent observers. Levels of independent observer coverage in fisheries vary depending on costs, value of the industry, political expedience and other factors.

Most fisheries in southern Australian waters have some interactions with seals. A review of these interactions was conducted by Peter Shaughnessy and others in 2003. Rather than re-review the interactions here, we describe selected examples.

Trawl fisheries

Trawl fisheries in Australia and New Zealand attract many fur seals, which feed on prey that are not normally available to them, and seals are often caught and drown in the nets, although many are also released alive (Plate 13). The South East Trawl Fishery operates in Commonwealth waters from just north of Sydney, around Victoria and Tasmania to Cape Jervis, South Australia, and is the largest scale-fish fishery in Australia. The primary method of fishing is single vessel, demersal trawling, although a small proportion of vessels use the Danish seine method. Between 1993 and 1995, two studies recorded rates of seal capture: Tony Harris and Peter Ward in a review of non-target catches in Australian waters reported 46 fur seals in 1886 trawls (2.4 seals per 100 trawls), and Peter Shaughnessy and Stephanie Davenport recorded three seals in 111 research trawls by CSIRO's research vessel *Soela* (2.7 seals per 100 trawls). During 1996 and 1997, fisheries researcher Ian Knuckey found that seals were caught in fewer than 1.5% of trawl shots, suggesting hundreds of seals drown each year.

Then, in winter months of 1999, 89 fur seals (six survived) were caught in 500 shots (about 18 seals per 100 trawls) conducted by factory trawlers off the west coast of Tasmania. Fisheries analyst Richard Tilzey suggested possible reasons for recording this large bycatch included the large size of trawlers and their fast towing speeds

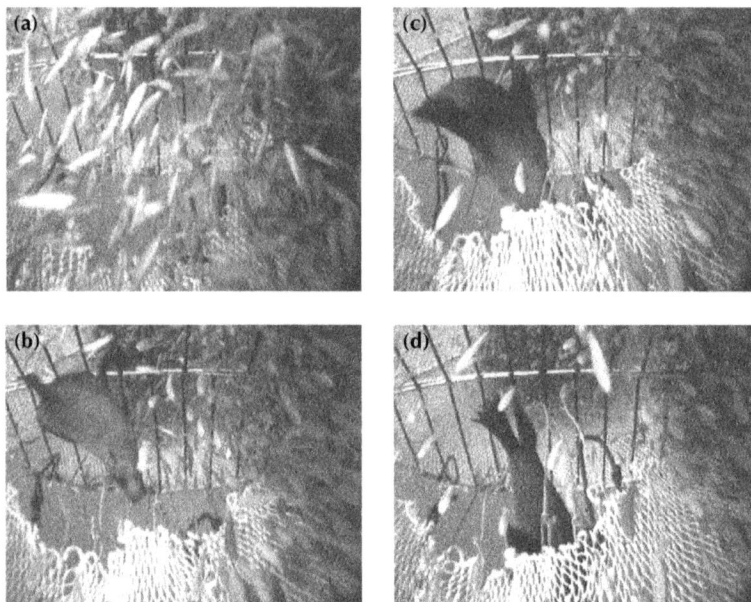

Figure 8.3. An Australian fur seal filmed as it caught fish in a trawl net and exited through a seal exclusion device fitted to reduce the chance seals would be caught and drowned.

Reproduced from Lyle and Willcox (2008) with permission J. M. Lyle, Institute for Marine and Antarctic Studies, University of Tasmania.

that year, as well as high observer coverage. The fishery had charted New Zealand factory trawlers, which were larger than trawlers normally used in Australian waters, to target a spawning aggregation of blue grenadier (*Macruronus novaezelandiae*). In the 2000 season, a seal exclusion device was trialled. A seal exclusion device is a grid inserted into the cod-end of the net that allows smaller items (including fish) to pass through while larger items (seals) are directed to the top (or bottom) where there is an escape hatch (Figure 8.3). Other strategies to reduce seal interactions included removing 'stickers', which are fish stuck in the mesh from the previous trawl, and reducing net-haul speeds (Figure 8.4). In 2000, 59 fur seals were caught in 467 shots (13%) and 22 of them died. All seals identified were Australian fur seals. Australian fur seals continue to be caught in trawl nets around southern Australia, although the numbers that drown are believed to be reducing as fishery codes of conduct regarding bycatch mitigation are adopted and new techniques are learned by fishermen to reduce interactions.

In New Zealand waters, New Zealand fur seals are frequently caught in trawl nets. A blue grenadier (hoki) fishery on the west coast of the South Island was estimated to have killed over 5600 New Zealand fur seals between 1989 and 1998 (1032 during the 1997–98 season). The 1999 season saw a substantial reduction in catches to an estimated 215 seals. Fur seals are also drowned in the nets of the southern blue whiting trawl fishery around the Bounty Islands, the Snares Island shelf squid fishery and the jack mackerel trawl fishery off the west coast of the South Island.

Bycatch of New Zealand sea lions in the commercial arrow squid trawl fishery near the Auckland Islands was first reported in 1978. Bycatch of between 17 and 143

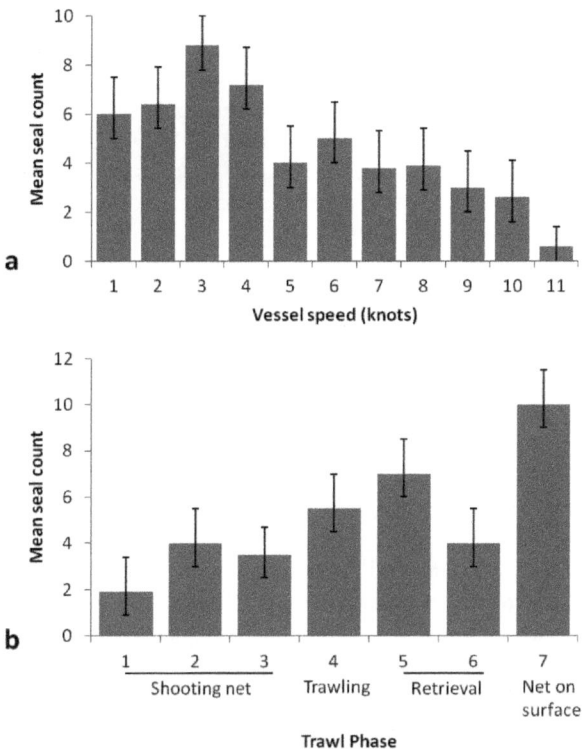

Figure 8.4. Numbers of Australian fur seals seen behind a trawl vessel in relation to (a) vessel speed and (b) stage in trawl process (when speeds are 2–4 knots). Derived from Hamer and Goldsworthy (2006). Reproduced with permission from Elsevier.

sea lions has been reported annually between 1995 and 2007, equating to 3.7% of mature seals in the population being caught annually, most of which (91%) are female. The population of New Zealand sea lions at the Auckland Islands declined by about 40% between 1998 and 2010 (12 years). The decline was linked to decreasing numbers of adult females returning to breed, with the principal cause of their demise being attributable to bycatch mortality in the squid fishery. Prominent strategies in place to manage New Zealand sea lion bycatch are fishery closures (12 nm closures around the Auckland Islands); a quota on the number of sea lions allowed to be caught each fishing season (when the catch of sea lions is exceeded, the fishery is closed for the remainder of the season); and use of SLEDs (sea lion exclusion devices).

Gill-net fisheries
Gill-net fisheries include commercial and recreational 'grab-all' nets used in bays and inlets, and in shark fisheries in shelf waters. They involve setting a wall of net which fish do not detect so swim into and are caught, often by the gills. Seals could be attracted to fish suspended in the nets and feed on them, thereby reducing the fisherman's potential catch. They also might become entangled in the nets and tear out sections, be cut free by the fishermen or drown. Individuals drowned in nets may fall out during net retrieval and thereby escape observation.

There are few quantitative data for gill-net interactions with seals. Tony Harris and Peter Ward noted that one Australian fur seal was taken in an experimental shark fishery that set 243 km of gill-nets in Bass Strait during the 1970s. Peter Shaughnessy reviewed some shark fishery logbook data. He noted that in 1998 one seal was recorded dead and two released alive from 14 243 shots and in 1999 one seal was recorded dead from 12 696 shots. In Victorian waters, fishing for shark is not permitted within 3 nautical miles of the coast and islands (fur seal breeding colonies), and fishermen generally believe they have few issues with seals.

One of the most significant bycatch situations in Australian fisheries occurs between the endangered Australian sea lion and the gill-net fishery for sharks in South Australia. Gill-nets were introduced to this shark fishery in 1963, peaked at 43 000 km of net-lifts in 1987 and plateaued at 15 000–20 000 km of net-lifts after 2000. Following substantial anecdotal evidence that interactions occurred, the first observer program to quantify the impact on the sea lions was conducted between 2004 and 2006. The results were presented by Simon Goldsworthy and co-workers in a report to the Fisheries Research and Development Corporation. This included results from an independent observer program undertaken by Derek Hamer over 146 sea days on 10 fishing trips, when 994 km of net-lifts and 12 sea lion mortalities were observed. As part of the study, a model of at-sea distribution and density of sea lions was developed based on 210 satellite tracked sea lions from 17 different colonies. A strong linear relationship between sea lion density and bycatch rate (per unit of fishing effort) was identified – that is, as the sea lion foraging density increased, so too did the probability of catching them. Applying this relationship to the whole fishery, it was estimated that over 200 sea lions were drowned annually in the fishery. Combined with other vulnerabilities of Australian sea lions (e.g. low reproductive rate, high philopatry), the rate of fishing bycatch mortality threatened the survival of many Australian sea lion subpopulations (Plate 14a). Based on the findings, the Australian Fisheries Management Authority, who managed the gill-net fishery, introduced spatial closures (most between 4–10 nm) around all sea lion colonies off South Australia in July 2010 and May and September 2011. They also introduce increased levels of independent observer coverage in the fishery (including onboard camera systems to monitor fishing activities) to improve monitoring of sea lion and other protected species interactions in the fishery, and a trigger limit system that places a cap on the total numbers of sea lions that are permitted to be caught within areas of the fishery, which if exceeded result in extended fishery closers. Additional management measures, such as switching gear to hook and line, are being considered.

Rock lobster fisheries

The fishery for southern rock lobster (*Jasus edwardsii*) off southern Australia commenced in 1968 and has averaged 2.3 million pot-sets per year. Seals can be attracted to the pots to take the bait, the rock lobsters or octopus (which enter the pots to eat the rock lobsters). They may also scavenge old baits or undersized lobsters that are discarded from the vessels. The interaction differs between species; fur seals tend to target the bait or octopus while sea lions can dexterously extract rock lobsters from pots.

The primary negative impact of this fishery on the seals is the capture and drowning of pups and juveniles. Bob Warneke reported that, in the 1970s, 43 of 182 recoveries (24%) of juvenile Australian fur seals tagged as pups at Seal Rocks were from animals that drowned in lobster pots. In Western Australia, Nick Gales and Richard Campbell separately documented mortalities of Australian sea lion pups and juveniles in western rock lobster (*Panulirus cygnus*) pots set adjacent to colonies (Plate 14b). Derek Hamer and others in South Australia found that whereas juvenile sea lions could be adept at removing rock lobsters, pups that entered pots often panicked and could not relocate the entrance. Unless removed immediately, they would drown.

Practical solutions, including tougher bait-holder designs to reduce bait take, and bars across the entrance or vertical spikes in the centre of the entrance to prevent seals from entering pots, can mitigate the interactions but were adopted slowly by the industry. This could be due to perceptions that rock lobster catch rates or sizes would also drop. A controlled research experiment in South Australia subsequently demonstrated that a steel upright (spike) attached to the base of the pot, and rising up to near the neck of the pot, virtually eliminated pot entry by sea lions and did not change the catch rate or size of lobsters.

Indirect fishery interactions

Indirect interactions between fisheries and seals are likely to be more substantial than direct interactions. They result from fisheries and seals targeting species within the same ecosystem and are often referred to as ecosystem or trophic interactions. In the most obvious examples, the same species and age classes of a prey are targeted by both consumers in the same area. This becomes more complicated when the same age classes of the same prey are targeted in different areas, or different age classes of the same species are targeted, or one consumer targets a species that is a predator or prey of a species targeted by the other consumer. Marine food webs are complex (Figure 8.5). Moreover, fisheries and seals target a range of species. Thus the effects of one on the other can be exceedingly convoluted and difficult to estimate.

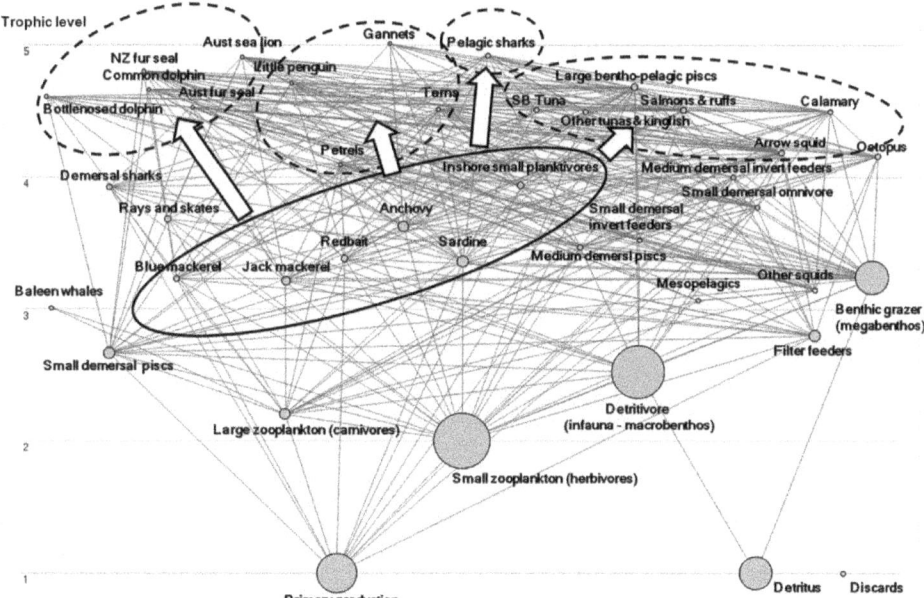

Figure 8.5. A partial food web developed for the Great Australian Bight ecosystem waters that includes Australian and New Zealand fur seals and Australian sea lions. From Goldsworthy *et al.* (2011).

There has been substantial debate over what degrees of trophic interactions between seals and fisheries are possible and to what extent intervention might mitigate them. Could seal predation seriously compromise the economic viability of a commercial fishery and could reductions in seal populations improve fishery viability? In Canada, the political opinion was 'yes, and something should be done about it'. This provided justification for the culling of harp seals, 'to make cod stocks more available to a fishery'. Could a fishery also be directly responsible for reducing the size and/or preventing recovery of a seal population? In the USA, the political opinion for this different situation was 'yes, and something should be done about it'. This led to area closures for lobster fisheries around the Hawaiian islands to protect endangered Hawaiian monk seals (*Monachus schauinslandi*), and for a pollock fishery in the eastern Bering Sea and Gulf of Alaska to protect endangered Steller sea lions (*Eumetopias jubatus*).

Such controversies highlight the need to better understand the role of seals in their marine ecosystems, as well as how commercial fisheries can have direct and indirect effects on other parts of the marine food web. In the early 2000s, a 'precautionary principle' was adopted by Australian fisheries. With respect to marine mammals, this principle meant that fishing operations needed to prove that they were exerting minimal impact, rather than waiting for evidence to prove an impact was occurring. This led to risk assessments for fishery and marine wildlife interactions, and a strengthening of attempts to quantify interactions.

In 2003, Simon Goldsworthy and co-authors collated available data to assess the extent of trophic interactions between otariids and fisheries in southern Australia. Their approach was to (1) develop population and bioenergetic models for the three otariids present and estimate annual prey consumption levels; (2) develop spatial models of distribution of foraging effort; (3) collate data on fisheries catches; and (4) model ecosystem interactions between the target species, their prey and their predators (which included seals and fisheries). The study represented a 'first attempt' which could be expanded as further data were collected. The authors detailed the many caveats of the study, but arrived at interesting conclusions. The seals were taking more prey than was harvested by the fisheries but generally were not competing directly with the fisheries for the same species. They took smaller bodied prey, mostly species that were not of high commercial significance, and they had few core foraging areas that overlapped spatially with fisheries. One of the most interesting results of this study was the mixed impact of increasing seal biomass on the biomass of commercially fished species. Model outcomes suggested that increasing seal biomass may reduce the available biomass of some fisheries and increase the biomass of others, but overall not impact on the total fish biomass available to fisheries production.

In research for his 2004 PhD thesis at Macquarie University, Charles Littnan modelled trophic interactions in eastern Victorian shelf waters, where there was high Australian fur seal foraging effort and a substantial commercial fishery. Using the same technique as Goldsworthy, Littnan also included potential biases in seal dietary analysis into his modelling. Despite variations in model outputs caused by the biases, the direction and magnitude of simulated biomass-change projections tended to stay the same for the majority of species modelled. It appeared that accuracy in population modelling and other model components were more vital than having completely accurate diet reconstructions.

As further data are collected on seal demographics, prey demographics and trophodynamics, and responses of species to changes within their ecosystem, the accuracy of fishery–seal–ecosystem models and simulations will improve.

Entanglement in marine debris

The debris that entangles seals in Australian waters is primarily derived from fishing operations. Seals may entangle themselves in fragments of nets, lines, ropes and box-straps that are dislodged or discarded during both commercial and recreational fishing activities. Material can also come from other sea craft (cargo-net, for example), terrestrial sources (including ribbons attached to party balloons, and plastics) and even occur naturally (such as sections of bull kelp). The debris may range in size from single loops to large sections of nets, weighing several kilograms (Plate 15).

Seals observed entangled in marine debris in Australian waters include Australian fur seals, New Zealand fur seals, subantarctic fur seals, Antarctic fur seals, Australian sea lions, leopard seals and southern elephant seals. In most instances, the seals become entangled around the neck. Some seals are also snared around the flippers and shoulders and others bite baited hooks and lures that catch in their mouths. Entanglement may lead to the death of the individual by increasing energetic demands, inhibiting effective foraging and/or cutting into the flesh. Observed incidences of entanglement underestimate the actual figure, because entangled individuals may die at sea prior to being detected.

In 1989, David Pemberton and others from Tasmanian Parks and Wildlife reported that 1.9% of Australian fur seals observed at haul-out sites in southern Tasmania were entangled in debris. This was one of the highest incidences of entanglement recorded for a marine mammal. The material seen most frequently on the seals was yellow packaging (box) straps that derived from bait boxes used in a foreign long-line vessels which operated in the region. After about 1998, the proportion of packaging straps entangling seals declined in association with the movement offshore of the fishery. A reverse of this trend was apparent at Marion Island, south Indian Ocean, where the incidence of entanglement of pinnipeds nearly doubled after 1996, in association with the arrival of long-line fisheries in the area.

At Seal Rocks between 1989 and 1994, Ron Prendergast from Melbourne Zoo observed that 0.8 to 1.2% of Australian fur seals were entangled and at Kanowna Island in 1996 it was 0.68%. Ron presented his findings in talks, reports and diving magazines over several years to maximise public awareness of his findings. Estimates of the incidences of entanglement for other otariids in southern Australian waters include 0.2% for Australian sea lions in 1989 (by Nick Gales), and 0.07% for New Zealand fur seals at several colonies on Kangaroo Island in 1994 and 1995 (by Peter Shaughnessy). Comparable values elsewhere are 0.15 to 0.58% for northern fur seals *Callorhinus ursinus* at St Paul Island, Alaska, between 1967 and 1996 and an average of 0.1% among juvenile Cape fur seals that were harvested in southern Africa during the 1970s.

In 2004, Brad Page and colleagues reviewed seal entanglements in Australian waters and estimated that over 1000 seals were dying each year as a result of marine debris. Rates of entanglement for Australian sea lions (1.3% in 2002) and New Zealand fur seals (0.9% in 2002) were the third and fourth highest reported for any otariid species. The sea lions were predominantly entangling in sections of monofilament gill-nets that originated from shark fisheries, while New Zealand fur seals were entangling in packaging straps (derived from rock lobster bait boxes) and trawl-net fragments (from local trawl fisheries).

Monitoring of Australian fur seals at the Seal Rocks colony between 1998 and 2012 recorded over 335 individually identifiable, entangled seals (Figure 8.6); 49% of these were caught and had the material removed. Entangled seals were seen at a rate of 2.5 per day spent at the colony. Trawl-net accounted for 40% of the material seen, although the prevalence of trawl-net declined over time. Monofilament fishing line often with associated hooks, lures, sinkers or floats (mostly from recreational fisheries) accounted for 24% of the entanglements. Most entangled seals (88%) were pups and juveniles, probably because their inquisitive and playful nature attracted them to drifting debris.

Most of the data on entanglement of seals in marine debris in Australian waters show incidences as great, or greater, than levels recorded elsewhere. It is generally presumed that most individuals are likely to die as a result of their entanglement, but the impacts of this mortality at the population level have rarely been evaluated. In northern fur seals, however, incidences of entanglement of up to 0.5% were implicated in population declines during the 1980s. In Australian waters, the small and genetically distinct populations of Australian sea lions are the most at risk.

Tourism

Several behavioural traits of pinnipeds lend them to tourism potential: species are generally colonial, providing a viewing spectacle; their annual attendance patterns at sites are predictable; and they exhibit interactive and 'playful' behaviours, which appeal to the public. Viewing experiences with pinnipeds range through guided

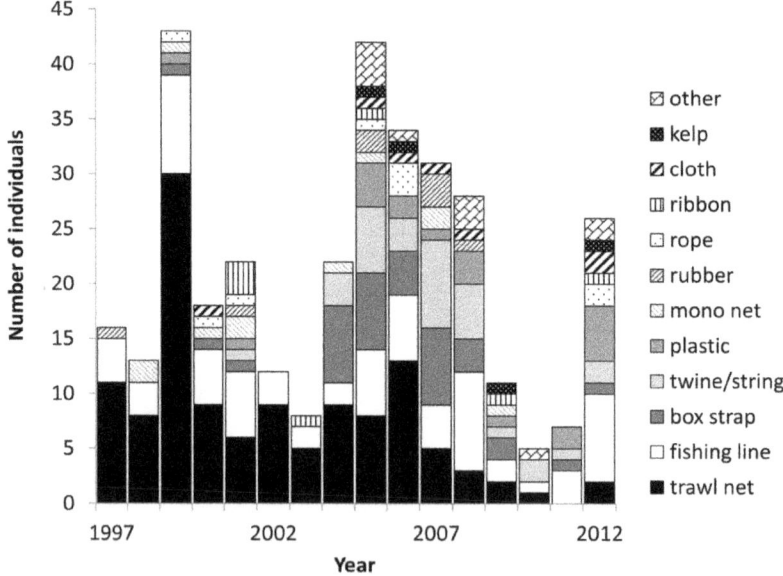

Figure 8.6. Entangled Australian fur seals recorded at Seal Rocks, Victoria, during near bimonthly visits, summarised by year and material. Roger Kirkwood unpublished data.

tours on shore, boat cruises, swimming and scuba diving interactions. Peak time is during summer, when the viewing spectacle of pinniped pupping coincides with public holidays. Seal-focused tourism in Australia increased rapidly during the 1990s (Plate 16).

A tourism industry around a wildlife attraction both indicates and stimulates public interest. It can be beneficial, increasing public desire to conserve wildlife, or detrimental, causing wildlife numbers to reduce. The seals can interpret humans as predators and flee, which can interrupt vital activities, increase energy expenditure and reduce breeding success. Codes of conduct (guidelines) to control tourist activity tend to develop only after some impact has occurred and in response to conflict between operators who report the bad practices of their competitors. Also, there is a tendency to introduce codes of conduct without enforcing compliance to them or monitoring their effectiveness. Pinniped-focused tourism probably has expanded in a somewhat unregulated fashion to fulfil consumer requirements.

State agencies are responsible for licensing and policing tourism industries that focus on pinnipeds. In Western Australia, for instance, licensing of commercial charters viewing pinnipeds commenced in 1995 and in 2001, 21 permits were issued to 15 licencees. In Victoria, marine mammal legislation (*Wildlife (Marine Mammals) Regulations 2009*) that included approaches to pinnipeds was introduced by authorities in 2009. These established a permitting system for pinniped viewing operations and restricted vessel approaches to breeding sites.

In 2003, a review of pinniped related tourism in the Southern Hemisphere was conducted by Roger Kirkwood and others. In Australia, tourism businesses have been established at most seal and sea lion sites that are accessible via day trips from coastal ports. Indeed, the main factor restricting pinniped-focused tourism appears to be accessibility to sites. Sustainable ventures are either close to human population centres, associated with a suite of attractions, such as dolphins, penguins, scenery or fishing opportunities, or offer unique experiences, such as 'swim-with' opportunities.

In Western Australia, the most visited pinniped sites are Carnac Island and Seal Island 11 km south-west of Fremantle, where sea- or shore-based viewing of non-breeding Australian sea lions is possible. In South Australia, the most popular site for viewing Australian sea lions is at Seal Bay on Kangaroo Island. Since the 1988, guided tours have been conducted at this site by operators that are accredited by National Parks and Wildlife, South Australia. Customers traverse boardwalks to a sandy beach and are guided around the sea lions. Visitation has increased from 20 000 per annum in the 1970s to 110 000 per annum by the late 2000s. Australian sea lions do not appear to be negatively influenced by the tourism. Cape du Couedic is another important pinniped viewing site on Kangaroo Island, and is one of the few locations in Australia where three pinniped species (Australian sea lions, Australian fur seals and New Zealand fur seals) can be seen together. At Baird Bay, on South Australia's Eyre Peninsula, an operator takes tourists to swim in shallow water with Australian sea lions at Jones Island, a breeding colony of about 70 individuals.

In Victoria, several pinniped sites are close to the capital city, Melbourne. At Phillip Island, 120 km south of Melbourne, visitors can go to the Seal Rocks colony of Australian fur seals in a 20 m, 200 passenger vessel, or view seals at the colony with the aid of a live video-link from cameras installed in a tower at the colony to a sea life centre on Phillip Island (Plate 16b). Seals also frequent man-made structures in Port Phillip Bay and charter boat tours commenced to these in the 1980s. The main component of the tours is the possibility of swimming with dolphins, but these are not always encountered whereas the seals are a reliable attraction. Another site regularly visited by charter boats since 1997 is Cape Bridgewater, a colony site for New Zealand and Australian fur seals. The Tasmanian Government reviewed the potential for seal tourism in the state and actively encourages seal viewing opportunities through the distribution of a 'where to see them' brochure. Boat-based operators offer tours to watch seals at about eight sites. In southern New South Wales, operators have provided seal viewing opportunities since 1984 at Montague Island, off Narooma, and Steamers Head, near Jervis Bay.

Guidelines (or 'codes of conduct') for approaches to marine mammals are designed for the protection of the viewers and to ensure viewers have minimal or no impact on the mammals. In most instances, guidelines are not established unless a problem is anticipated: few guidelines are totally precautionary. Perhaps the principal factor influencing the establishment of guidelines is prevailing social culture. Often guidelines are prepared by tour operators. An enhancement of protection comes when governments, acting in the public interest, introduce regulations. Critical factors influencing the effectiveness of regulations are enforcement capability and research. Regulations will not be adhered to if there is no enforcement capability and may be argued against if there is no research to support their application. In addition, operators should be well educated in pinniped behaviour and conservative in their approach, or may develop habits that compromise pinnipeds and tourists, reducing the sustainability of the industry.

When on land, seals perceive threats by sight, sound and smell, and generally respond by becoming alert (changing posture and directing attention towards the threat) and moving to the water. In severe cases, disturbances can elicit stampedes involving hundreds of animals and, due to the rough terrain of some sites, can result in injury or death. Young pups are especially at risk of injury during stampedes. In addition to their vulnerability to being trampled, they are not yet fully adapted to marine life. In the water, they are at much higher risk of predation, drowning and hypothermia than other age groups. Disturbances at a breeding site may delay a female's return, disrupt suckling and cause a female to depart prematurely, all of which can reduce maternal energy transfer to the pup and therefore lower the likelihood of its survival.

Laura Boren, working at several sites in New Zealand, found New Zealand fur seals responded strongly to land-based approaches within 10 m by heightening vigilance, changing posture, and/or fleeing, while student Jean-Paul Orsini found approaches to the same distance only increased vigilance in Australian sea lions. Terri-Jo Lovasz and colleagues investigated thresholds of Australian sea lions at Seal Bay to tourism pressure. Her study found that some Australian sea lions reacted to the presence of people that were 30 m away and that sea lions in areas that were frequented by tour groups were less likely to react to the approach of people than sea lions that were resting in areas not frequented by tourists. One recommendation that was adopted from her study included limiting the approach distance of tour groups (in areas used by tour groups) to sea lions to 10 m, replacing the former limit of 6 m.

Peter Shaughnessy and others investigated the responses of Australian and New Zealand fur seals to boat-based stimuli at haul-outs on Montague Island, NSW, using both controlled approaches and observations of routine boat traffic. Behavioural responses observed were similar in both species, and included increased vigilance and fleeing behaviours, but Australian fur seals were significantly more responsive than New Zealand fur seals. Alex Burleigh and co-workers observed comparable responses at a haul-out at Steamers Head, New South Wales. Haul-outs consist mainly of adult male and juvenile seals, while natal colonies are typically more densely populated and primarily occupied by females and pups, age classes that are significantly more sensitive to disturbance.

Masters student Julia Back found Australian fur seals at colonies showed a broad range of responses to boat approaches, the severity of which were dependent on approach distance, time of day, age/gender class, distance to water, site characteristics and previous exposure to boats. Boat approaches at a colony that was relatively naive to boat traffic (Kanowna Island) reduced resting and suckling behaviours and many seals fled into the water. Seals were more responsive to approaches during the morning than the afternoon, and more responsive to closer approaches (25 m) than to more distant approaches (75 m). Pup responses varied through the year: during the summer post-breeding period, the majority remained ashore, while in the winter post-moult period most fled into the water. Suckling behaviours were affected most strongly during the summer post-breeding period, when pups are small and appear to be most vulnerable. The observed frequency of fleeing behaviours suggests that Australian fur seals at Kanowna Island perceived boat approach as a substantial threat to their survival, and that the precautionary protocols developed from anecdotal evidence for this site are ineffective at protecting seals during the winter post-moult period and may be excessively limiting to ecotourist operators during the summer post-breeding period.

At a colony exposed to considerable boat traffic (Seal Rocks), Back found seals were considerably less sensitive to routine tourist boat approaches than seals at the naive colony. Though there were consistent small increases in activity and decreases in attendance (under 2%), a large-scale response was observed only once, on the return of the tour vessel after a 2-month absence. The results suggested that routine exposure to boats may increase the Australian fur seals' tolerance of them, provided the visits are frequent and consistent. Moreover, management guidelines should be based on quantified, species-specific data that takes into account individual site characteristics.

Toxins

Toxins can damage the immune, endocrine and nervous systems of marine mammals, disrupting growth and resistance to disease, and causing individual and mass mortalities. The toxins may be caused or enhanced by natural events, such as algal blooms and geological processes, or by anthropogenic sources.

Algal blooms are rapid accumulations of one or more species of phytoplankton that result from abrupt increases in light or nutrients. Not all algal blooms are harmful. Those that can become harmful to pinnipeds do so by either killing their prey, through mechanical damage to gills or water eutrophication (removal of oxygen through degradation of dead cells), or by producing natural toxins that bio-concentrate up the food chain. For example, high concentrations of domoic acid in waters off California result from blooms of the algae *Pseudonitzchia australis* and have led to seizures, behavioural abnormalities and mortalities in Californian sea lions. Algal blooms have not been linked to pinniped mortalities in Australia as yet, but were considered as a causative factor in a mass mortality of New Zealand sea lions at the Auckland Islands in 1998.

There are thousands of human-produced or enhanced substances that could enter the marine environment and be toxic to marine animals. They can be broadly divided into chemical compounds, including organochlorines (e.g. pesticides), hydrocarbons (e.g. oil) and metals. Organochlorines typically are resistant to degradation and readily accumulate in marine food chains, concentrating in top level predators, such as otariids. Hydrocarbons, including fuel lost from ships, may break down rapidly and be avoided by otariids at sea, but if washed ashore at colonies and haul-outs they cannot be avoided. Some metals, like copper, zinc and iron, are essential for normal functioning of an animal but may be toxic in high concentrations, while others such as mercury, cadmium and lead are non-essential. Non-essential metals may affect body processes and immune functions even at low concentrations.

There have been few studies of pollutant exposure and levels in Australian otariids. One study of organochlorine levels in a range of marine mammals by Alex Aguilar from the University of Barcelona and co-workers compared levels of DDT (dichloro-diphenyl-trichloro-ethane, a synthetic pesticide) and its break-down product DDE (dichloro-diphenyl-dichloro-ethylene) in small numbers of Australian and Cape fur seals. Juvenile and adult male Australian fur seals contained higher concentrations of most compounds while the distribution of pollutants in adult females varied. Within Australian fur seals, higher concentrations occurred in males than females. This pattern is commonly observed in marine mammals. However, Cape fur seal females contained higher concentrations than males. The small sample sizes prohibited more definitive discussion of the different accumulation patterns. Other studies have noted that DDT, PCB (polychlorinated biphenyl, a synthetic organic compound with many applications) and mercury levels are low in Australian fur seals compared with in otariids from comparable temperate regions of the Northern Hemisphere, but higher than in high-latitude phocids. However, Michael Lynch did find high pollutant levels in Australian fur seals at Lady Julia Percy Island (see the section on diseases in the previous chapter). The different levels of pollutants among seals probably relate to each seal's relative proximity to industrialised areas.

Oil spills along coastlines may result in otariid deaths by reducing the insulation properties of the fur, through inhalation of fumes, and through grooming and ingestion of toxic concentrations. In February 1997, an oil tanker lost up to 5000 tons of crude oil off the coast of Uruguay, part of which washed ashore on Isla de Lobos, an island colonised by South American fur seals and Southern sea lions. Movement of the seals spread the oil over much of the island. The oil killed 5000–10 000 fur seal pups (from an annual production of approximately 30 000), and an unrecorded number of other fur seals and sea lions. The actual cause of death, ingestion or otherwise, was not investigated. Perhaps the world's most famous oil spill to date was from the *Exxon Valdez* in Prince William Sound, Alaska, which killed several hundred harbour seals. In Australia, two oil spills have affected seal populations: the 1991 *Sanko Harvest* spill in Western Australia oiled almost 200 New Zealand fur seals pups (which were successfully cleaned) at two colonies and the 1995 *Iron Barron* spill in Tasmania affected an unknown number of Australian fur seals at the Tenth Island colony.

Lone seal management

Each year, hundreds of thousands of seals come ashore at recognised colony and haul-out sites around the southern Australian coastline. On hundreds of occasions each

Figure 8.7. Lone seals may come ashore at unusual places for a variety of reasons, including (a) too young to be at sea (washed away from their colony); (b) unwell, resting or moulting; (c) curiosity or misadventure. Photographers: (a, b) Roger Kirkwood; (c) Ken Mankey, Port Welshpool.

year, seals also come ashore singly at other locations and are appreciated by human observers as being 'interesting'. Live seals may come ashore for a variety of reasons including tiredness or exhaustion, injury, disease or malnourishment (Figure 8.7). Tired seals need to rest on the beach for several hours or up to several days. Entangled seals may require quick intervention to remove the debris before they return to the water. Seals can recover from seemingly major injuries and intervention could be a greater stress to them than recovering normally in the wild. A few injured or diseased seals may warrant assistance, in the form of mercy euthanasia, to provide relief from suffering. This is something for a trained veterinarian or wildlife officer to decide. Most seals that present to the public on beaches and instil a desire to 'do something' about them are malnourished. Starvation is perhaps the most common form of mortality for any wildlife, and starving seals frequently come ashore to rest on beaches where they may be seen by humans. The starving individual provides humans with the greatest conundrum: should the seal be left alone, euthanased or taken into captivity and 'rehabilitated'?

Whether seals should be taken into captivity requires considered judgment. Appropriate facilities with space and veterinary support need to be available. Seals taken into captivity quickly habituate to humans and when released may fail to readapt to the wild, or become 'problem' seals. Pups taken into captivity need to be bottle fed for an extended period, taught to capture fish and taught not to expect fish from humans prior to release, which is a difficult task. If a seal is taken into captivity, even if the initial intention is to hold it just for several weeks, it must be recognised it may never be 'releasable'. It may need to be held in captivity for the rest of its life (up to

20 years). This can be expensive, labour intensive and not what many people would consider a good life for the seal. If a seal is non-releasable and there is no place for it to go, it would have to be euthanased. Another consideration when contemplating whether a seal should be taken into captivity is: could it spread disease to other seals, either in the captive facility or to the wild population if released? In a holding facility in Western Australia during the 1980s, a tuberculosis virus in captive New Zealand fur seals spread between all the seals in the facility (all seals were euthanased) and was passed onto a seal handler and to seals in another facility.

If taking beached seals into captivity is not an option, should we euthanase those we suspect are about to die? If in the opinion of a qualified wildlife officer or veterinarian, the seal is unduly suffering from an obvious gross mutilation or disease from which it cannot recover, euthanasia may be an option. However, if the seal is simply starving, should we intervene? Is this seal suffering unduly, when starvation is the most common form of mortality amongst seals, and tens of thousands of seals die from it each year? Some people will feel strongly that 'yes, all animals should be assisted if we can do so'; others will feel 'no, it is a natural process and therefore we should not intervene'.

Motivation to intervene with a seal on a beach may be stimulated by public pressure. If large numbers of people are disturbing an 'unwell' seal, its chance to recover could be compromised and the harassment to it may be seen as inhumane. Then wildlife officers may consider it appropriate to euthanase the seal, establish barricades around it or translocate it to a 'quieter' location.

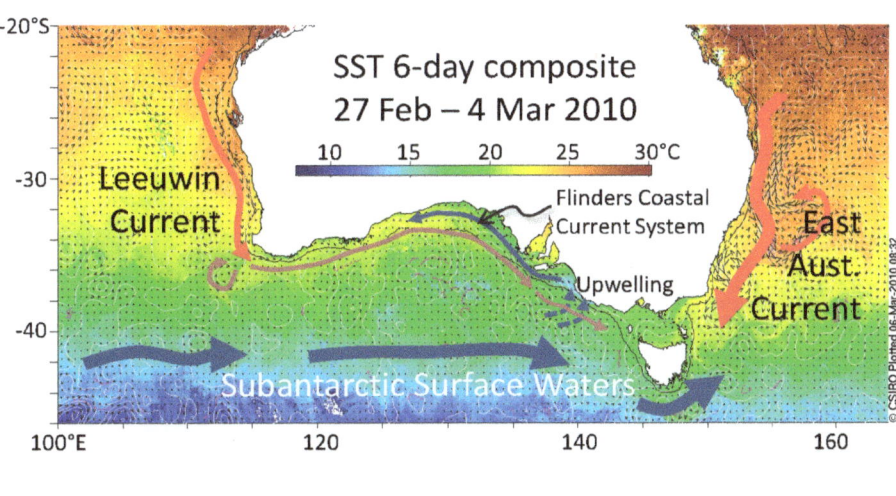

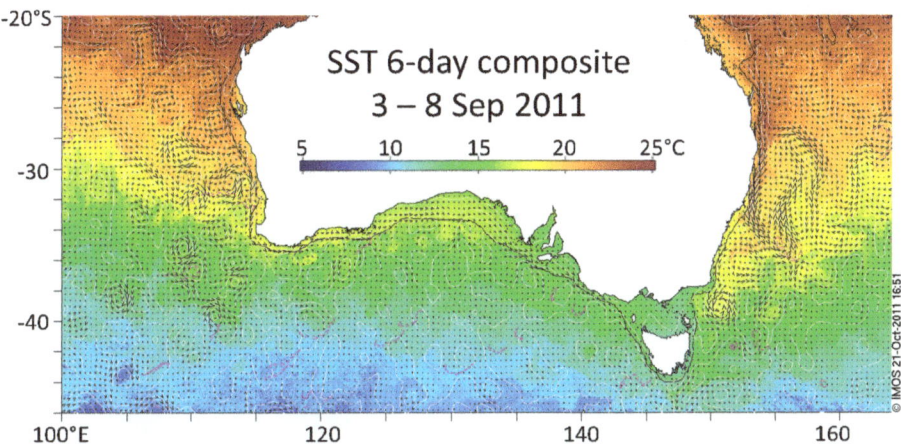

Plate 1a. Surface ocean currents and representative autumn and spring sea-surface temperature images of Southern Australia. Images courtesy CSIRO & IMOS (webpage http://oceancurrent. imos.org.au/sst_s/).

Plate 1b. Many sites occupied by seals in Australia are low-lying and may be inundated by high seas. Photographer: Roger Kirkwood.

Plate 2a. Australian fur seals exhibiting thigmotaxis, touching each other, which is not displayed so strongly by New Zealand fur seals. Photographer: Roger Kirkwood.

Plate 2b. External ears of the Australian fur seal. Photographer: Roger Kirkwood.

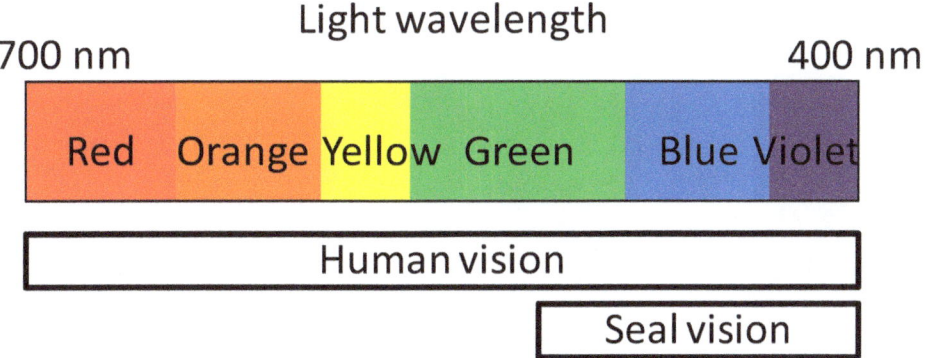

Plate 3a. Comparison of colour vision in humans and seals. Sourced from Berta and Sumich (1999), copyright Elsevier, reproduced with permission.

Plate 3b. Australian fur seals resting and sleeping at sea during calm weather. A foreflipper is often held in the air for balance and to detect a breeze or for cooling. Photographer: Roger Kirkwood.

Plate 4. Playful behaviours of young Australian fur seals include wave riding, leaping, chasing and toying with objects.
Photographers: (a, c, d) Roger Kirkwood; (b) Vincent Antony; (e, f) Jay Town, *Herald Sun Melbourne*, permission sourced from Newspix.com.au.

Plate 5. Comparisons of different age classes of the three otariids that breed in southern Australia. Photographers: (Australian sea lion) Brad Page, Heidi Ahonen and Simon Goldsworthy; (New Zealand and Australian fur seals) Roger Kirkwood.

Plate 6. Australian fur seal pups demonstrating the change in colour from birth to post-moult. After moult the pups are difficult to distinguish from juveniles: (a, b) December–January; (c, d) March–April; and (e, f) May–June. Photographers: (a, c, d, e, f) Roger Kirkwood; (b) James Archibald.

Plate 7. Seals that may visit the coastline of continental and subantarctic Australia: (a) elephant seal; (b) leopard seal; (c) Weddell seal; (d) crabeater seal; (e) Ross seal; (f) subantarctic fur seal; (g) Antarctic fur seal; (h) New Zealand sea lion. Photographers: (a, f) Simon Goldsworthy; (b, c, h) Roger Kirkwood; (d, g) John Kirkwood; (e) © Australian Antarctic Division.

Plate 8a. Territories are maintained largely through posturing; however, fights between similarly sized adult males occur occasionally, particularly during territory establishment.

Plate 8b. Birth of an Australian fur seal pup. Photographer: John Gibbens.

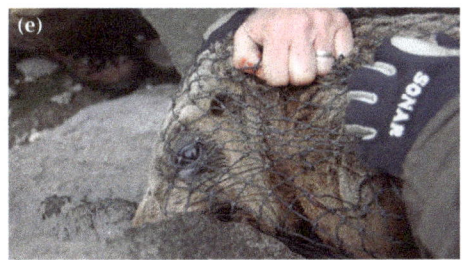

Plate 9. Methods for seal capture and restraint vary depending on the size of the seal, the terrain and the purpose for capture: (a) picking up a pup by the hindflippers; (b) netting a flighty Australian fur seal (c) netting a 'stand-and-fight' Australian sea lion; (d) darting a large or flighty seal; (e) temporary restraint to remove an entanglement; (f) dip-netting a seal at sea; and (g) anaesthetising a seal when restraining for an extended period (more than 10 minutes).

Photographers: (a, b) Tony Mitchell; (c, d) Simon Goldsworthy; (e, f, g) Roger Kirkwood.

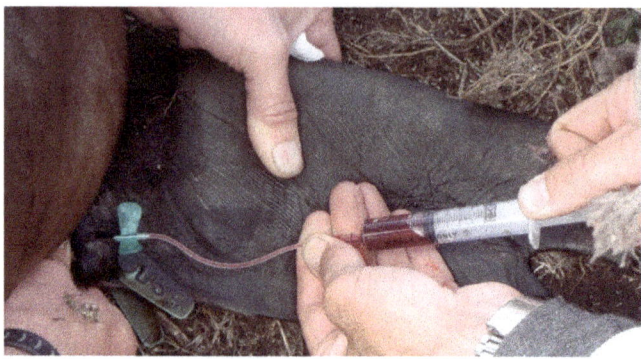

Plate 10a. Taking a blood sample from the pectoral vein of an Australian fur seal. Photographer: Roger Kirkwood.

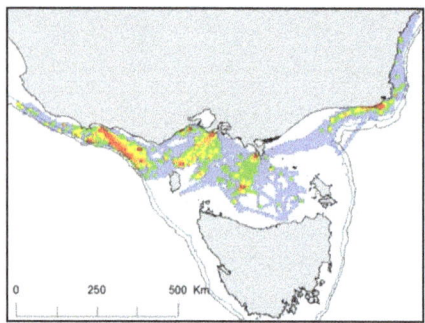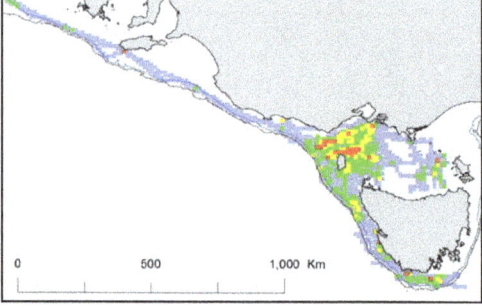

Plate 10b. Australian fur seal time-in-area analysis, red represents greatest densities at sea for (a) lactating females from four colonies in Victoria; and (b) adult males from the Seal Rocks, Victoria. Data published in Kirkwood and Arnould (2011), and Kirkwood, Lynch et al. (2006). © 2008 Canadian Science Publishing or its licensors. Reproduced with permission.

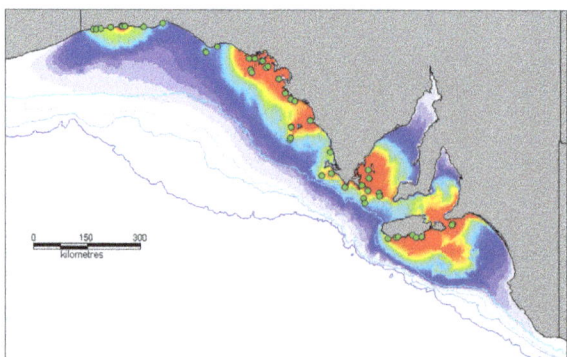

Plate 10c. Foraging ranges of Australian sea lions from colonies in South Australia. Reproduced from Goldsworthy et al. (2011).

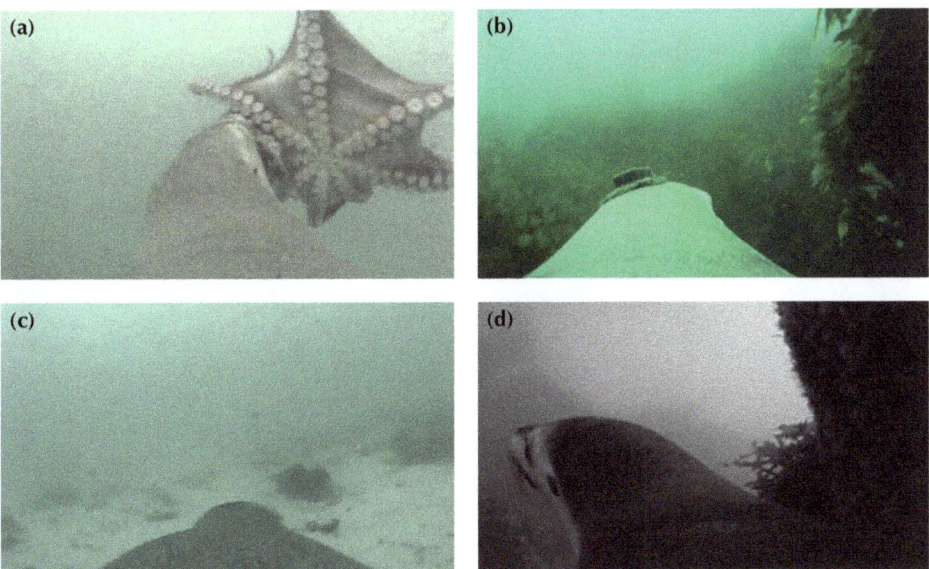

Plate 11a. Crittercam photos from on-board a female Australian sea lion: (a) eating an octopus; (b, c, d) different foraging habitats. Photographer: Simon Goldsworthy. Photographs courtesy National Geographic Remote Imaging.

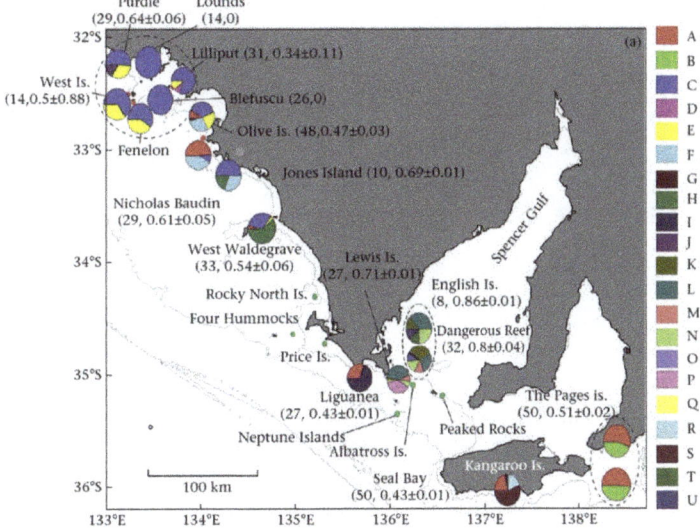

Plate 11b. DNA haplotype structure in Australian sea lion populations in South Australia demonstrating a high level of separation between sites. Reproduced from Lowther, Harcourt *et al.* (2012), with permission from Elsevier.

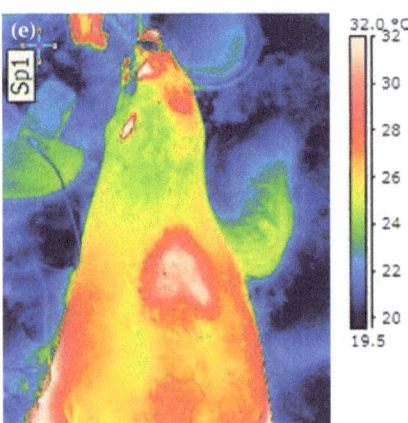

Plate 12. Alopecia (hair loss) in Australian fur seals is prevalent amongst juvenile females at Lady Julia Percy Island: (a) mild case; (b) severe case; (c) very severe case; (d) a group of seals, several of which suffer alopecia; (e) thermal image indicating inability to control heat exchange across skin where there is hair loss (white colour indicates highest temperature).

Photographers: (a, b, c, d) Roger Kirkwood; (e) Michael Lynch.

Plate 13. Australian fur seals are caught in trawl nets in southern Australian waters. Some seals are released alive, others drown: (a) a factory trawler; (b) releasing a seal from a trawl net, (c) encouraging a seal to leave the ship via the trawl ramp. Photographer: Roger Kirkwood.

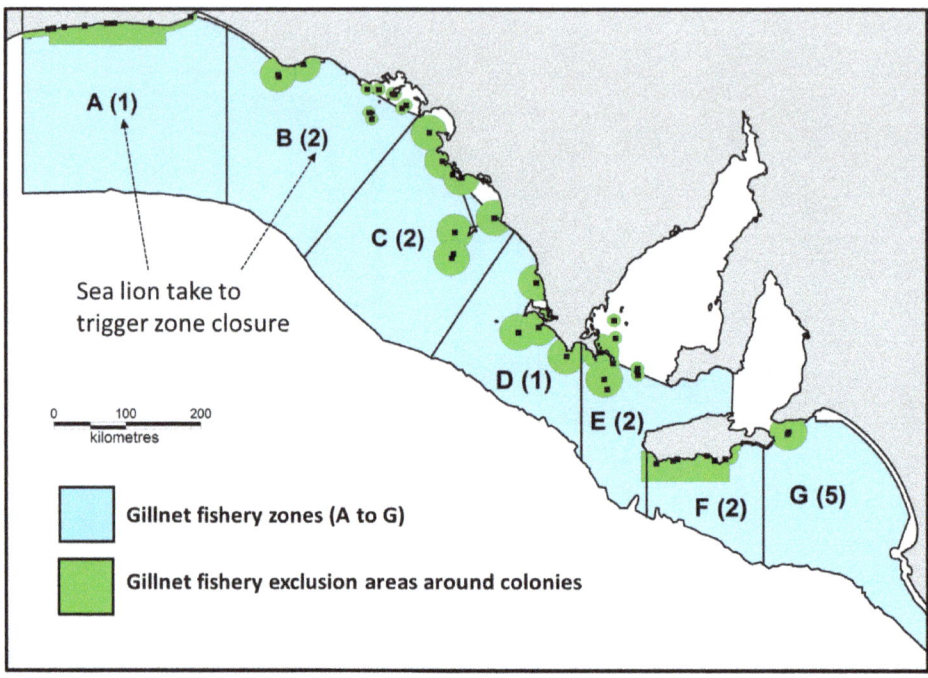

Plate 14a. Map detailing Australian sea lion bycatch management strategies introduced by the Australian Fisheries Management Authority (AFMA) in the shark gillnet fishery off South Australia in January 2012. Areas around colonies (green) are closed and fishing zones (blue, A to G) have annual sea lion bycatch limits (in brackets), which if reached trigger closure of the zone to gillnet fishing for 18 months. Map redrawn with permission from AFMA.

Plate 14b. Juvenile Australian sea lion removing a rock lobster from an experimental pot.
Photographer: Richard Campbell.

Colour plates

Plate 15. Examples of entangled seals: (a) orange net; (b) fishing line with float; (c) yellow rope; (d) fishing lure and hooks; (e) monofilament net; (f) trawl net; (g) unknown; (h) polypropylene line. Photographers: (a, e) John Gibbens; (b, c) Roger Kirkwood; (d) Warren Reed; (f) Julia Back; (g) Fran Solly; (h) Simon Goldsworthy.

Plate 16. Tourism: (a, b) boat at Seal Rocks; (c) Chinaman's Hat in Port Phillip Bay; (d) on the beach at Seal Bay, Kangaroo Island. Photographer: Roger Kirkwood.

BIBLIOGRAPHY

Aguilar A, Borrell A, Reijnders PJH (2002) Geographical and temporal variation in levels of organochlorine contaminants in marine mammals. *Marine Environmental Research* **53**, 425–452.

Arnason U, Gullberg A, Janke A, Kullberg M, Lehman N, Petrov EA, Väinölä R (2006) Pinniped phylogeny and a new hypothesis for their origin and dispersal. *Molecular Phylogenetics and Evolution* **41**, 345–354.

Arnould JPY, Boyd IL, Warneke RM (2003) The historical dynamics of the Australian fur seal population: evidence of regulation by man? *Canadian Journal of Zoology* **81**, 1–9.

Arnould JPY, Cherel Y, Gibbens J, White JG, Littnan CL (2011) Stable isotopes reveal inter-annual and inter-individual variation in the diet of female Australian fur seals. *Marine Ecology Progress Series* **422**, 291–302.

Arnould JPY, Costa DP (2006) Sea lions in drag, fur seals incognito: insights from the otariid deviants. In 'Sea lions of the world.' (Eds AW Trites, SK Atkinson, DP DeMaster, LW Fritz, TS Gelatt, LD Rea and KM Wynne) pp. 309–324. (University of Alaska Press: Anchorage)

Arnould JPY, Hindell MA (2001) Dive behaviour, foraging locations, and maternal-attendance patterns of Australian fur seals (*Arctocephalus pusillus doriferus*). *Canadian Journal of Zoology* **79**(1), 35–48.

Arnould JPY, Hindell MA (2002) Milk consumption, body composition and pre-weaning growth rates of Australian fur seal (*Arctocephalus pusillus doriferus*) pups. *Journal of Zoology* **256**, 351–359.

Arnould JPY, Kirkwood R (2008) Habitat selection by female Australian fur seals (*Arctocephalus pusillus doriferus*). *Aquatic Conservation: Marine and Freshwater Ecosystems* **17**, S53–S67.

Arnould JPY, Littnan CL (2000) Pup production and breeding areas of Australian fur seals *Arctocephalus pusillus doriferus* at Kanowna Island and The Skerries in north-eastern Bass Strait. *Australian Mammalogy* **22**, 51–55.

Arnould JPY, Littnan CL, Lento GM (2000) First contemporary record of New Zealand fur seals *Arctocephalus forsteri* breeding in Bass Strait. *Australian Mammalogy* **22**, 57–62.

Arnould JPY, Trinder DM, McKinley CP (2003) Interactions between fur seals and a squid jig fishery in southern Australia. *Marine and Freshwater Research* **54**, 979–984.

Arnould JPY, Warneke RM (2002) Growth and condition in Australian fur seals (*Arctocephalus pusillus doriferus*) (Carnivora: Pinnipedia). *Australian Journal of Zoology* **50**(1), 53–66.

Back J (2010) 'Behavioural responses of Australian fur seals to boat-based ecotourism.' (Deakin University, Burwood, Victoria)

Bax NJ, Williams A (2001) Seabed habitat on the south-eastern Australian continental shelf: context, vulnerability and monitoring. *Marine and Freshwater Research* **52**, 491–512.

Baylis AMM, Page B, Goldsworthy SD (2008) Colony-specific foraging areas of lactating New Zealand fur seals. *Marine Ecology Progress Series* **361**, 279–290.

Baylis AMM, Page B, Peters K, McIntosh R, McKenzie J, Goldsworthy S (2005) The ontogeny of diving behaviour in New Zealand fur seal pups (*Arctocephalus forsteri*). *Canadian Journal of Zoology* **83**, 1149–1161.

Bejder L, Samuels A, Whitehead H, Finn H, Allen S (2009) Impact assessment research: use and misuse of habituation, sensitisation and tolerance in describing wildlife responses to anthropogenic stimuli. *Marine Ecology Progress Series* **395**, 177–185.

Berry O, Spiller LC, Campbell RA, Hitchen Y, Kennington WJ (2012) Population recovery of the New Zealand fur seal in southern Australia: a molecular DNA analysis. *Journal of Mammalogy* **93**(2), 482–490.

Berta A, Churchill M (2011) Pinniped taxonomy: review of currently recognized species and subspecies, and evidence used for their description. *Mammal Review* **42**(3), 207–234.

Berta A, Ray CE (1990) Skeletal morphology and locomotor capabilities of the archaic pinniped *Enaliarctos mealsi*. *Journal of Vertebrate Paleontology* **10**(2), 141–157.

Berta A, Sumich JL (1999) 'Marine mammals: evolutionary biology.' (Academic Press: San Diego)

Biuw M, McConnell B, Bradshaw CJA, Burton H, Fedak MA (2003) Blubber and buoyancy: monitoring the body condition of free-ranging seals using simple dive characteristics. *Journal of Experimental Biology* **206**, 3405–3423.

Bodley KB, Mercer JR, Bryden MM (1999) Rate of passage of digesta through the alimentary tract of the New Zealand fur seal (*Arctocephalus forsteri*) and the Australian sea lion (*Neophoca cinerea*) (Carnivora: Otariidae). *Australian Journal of Zoology* **47**(2), 193–198.

Boren LJ, Gemmell NJ, Barton KJ (2002) Tourist disturbance on New Zealand fur seals, *Arctocephalus forsteri*. *Australian Mammalogy* **24**, 85–96.

Boyd I, Lunn N, Barton T (1991) Time budgets and foraging characteristics of lactating Antarctic fur seals. *Journal of Animal Ecology* **60**, 577–592.

Boyd IL, Bowen WD, Iverson SJ (2010) 'Marine mammal ecology and conservation: a handbook of techniques.' (Oxford University Press: Oxford)

Boyd IL, Lockyer C, Marsh HD (1999) Reproduction in marine mammals. In 'Biology of marine mammals.' (Eds JE Reynolds and SA Rommel) pp. 218–286. (Melbourne University Press: Melbourne)

Bradshaw CJA, Hindell MA, Littnan C, Harcourt RG (2006) Determining marine movements of Australasian pinnipeds. In 'Evolution and biogeography of Australian vertebrates.' (Eds JR Merrrick, M Archer, GM Hickey and MSY Lee) pp. 889–911. (Auscipub PL: Oatlands, NSW)

Bradshaw CJA, Lalas C, Thompson CM (2000) Clustering of colonies in an expanding population of New Zealand fur seals (*Arctocephalus forsteri*). *Journal of Zoology* **250**, 105–112.

Brothers N, Pemberton D (1990) Status of Australian and New Zealand fur seals on Maatsuyker Island, southwestern Tasmania. *Wildlife Research* **17**, 563–569.

Brunner S (1998) Cranial morphometrics of the southern fur seals *Arctocephalus forsteri* and *A. pusillus* (Carnivora: Otariidae). *Australian Journal of Zoology* **46**(1), 67–108.

Brunner S (2004) Fur seals and sea lions (Otariidae): identification of species and taxonomic review. *Systematics and Biodiversity* **1**(3), 339–439.

Brunner S, Bryden MM, Shaughnessy PD (2004) Cranial ontogeny of otariid seals. *Systematics and Biodiversity* **2**(1), 83–110.

Bunce A, Norman FI, Brothers N, Gales R (2002) Long-term trends in the Australian gannet (*Morus serrator*) population in Australia: the effect of climate change and commercial fisheries. *Marine Biology* **141**, 263–269.

Burleigh A, Lynch T, Rogers T (2008) Best practice techniques for monitoring the fur seal haul-out site at Steamers Head, NSW, Australia. In 'Too close for comfort: contentious issues in human-wildlife encounters.' (Eds D Lunney, A Munn and W Meikle) pp. 233–245. (Royal Zoological Society of New South Wales: Mosman, NSW)

Campbell R, Holley D, Christianopoulos D, Caputi N, Gales N (2008) Mitigation of incidental mortality of Australian sea lions in the west coast rock lobster fishery. *Endangered Species Research* **5**, 345–358.

Campbell RA, Gales NJ, Lento GM, Baker CS (2008) Islands in the sea: extreme female natal site fidelity in the Australian sea lion, *Neophoca cinerea*. *Biology Letters* **4**, 139–142.

Cane KN, Arnould JPY, Nicholas KR (2005) Characterisation of proteins in the milk of fur seals. *Comparative Biochemistry and Physiology. Part B, Biochemistry & Molecular Biology* **141**(1), 111–120.

Charrier I, Mathevon N, Jouventin P (2001) Mother's voice recognition by seal pups. Newborns need to learn their mother's call before she can take off on a fishing trip. *Nature* **412**, 873.

Charrier I, Mathevon N, Jouventin P (2003) Vocal signature recognition of mothers by fur seal pups. *Animal Behaviour* **65**, 543–550.

Crawford CM, Edgar GJ, Cresswell G (2000) The Tasmanian region. In 'Seas at the millennium.' (Eds C Shephard and LP Zann) pp. 647–660. (Pergamon: Amsterdam)

Cumpston JS (1970) 'Kangaroo Island, 1800–1836.' (Union Offset Company PL Fyshwick, ACT)

Cumpston JS (1973) 'First visitors to Bass Strait.' (Union Offset Company PL: Fyshwick, ACT)

Daniel JC (1981) Delayed implantation in the northern fur seal (*Callorhinus ursinus*) and other pinnipeds. *Journal of Reproduction and Fertility. Supplement* **29**, 35–50.

David J, Wickens P (2003) Management of cape fur seals and fisheries in South Africa. In 'Marine mammals: fisheries, tourism and management issues.' (Eds N Gales, M Hindell and R Kirkwood) pp. 116–136. (CSIRO Publishing: Melbourne)

Deagle BE, Kirkwood R, Jarman SN (2009) Evaluating trophic links by pyrosequencing prey DNA in faeces: a population level dietary study on Australian fur seals. *Molecular Ecology* **18**, 2022–2038.

De Graaf AS, Shaughnessy PD, McCully RM, Verster A (1980) Occurrence of *Taenia solium* in a Cape fur seal (*Arctocephalus pusillus*). *The Onderstepoort Journal of Veterinary Research* **47**, 119–120.

Dobson FS, Jouventin P (2003) How mothers find their pups in a colony of Antarctic fur seals. *Behavioural Processes* **61**, 77–85.

Edgar GJ, Samson CR, Barrett NS (2004) Species extinction in the marine environment: Tasmania as a regional example of overlooked losses in biodiversity. *Conservation Biology* **19**(4), 1294–1300.

Elsner R (1999) Living in water: solutions to physiological problems. In 'Biology of marine mammals.' (Eds JE Reynolds and SA Rommel) pp. 73–116. (Melbourne University Press: Melbourne)

Elsner R, Gooden B (1983) 'Diving and asphyxia: a comparitive study of animals and man.' (Cambridge University Press: New York)

Evans K, Thresher R, Warneke RM, Bradshaw CJA, Pook M, Thiele D, Hindell MA (2005) Periodic variability in cetecean strandings: links to large-scale climate events. *Biology Letters* **1**, 147–150.

Evans SR, Middleton JF (1998) A regional model of shelf circulation near Bass Strait: a new upwelling mechanism. *Journal of Physical Oceanography* **28**, 1439–1457.

Fish FE (1993) Influence of hydrodynamic design and propulsive mode on mammalian swimming energetics. *Australian Journal of Zoology* **42**, 79–101.

Flinders M (1814) 'A voyage to Terra Australis; undertaken for the purpose of completing the discovery of that vast country, and prosecuted in the years 1801, 1802 and 1803, in His Majesties' Ship the Investigator.' (G. and W. Nicol: London)

Fowler CW (1987) Marine debris and northern fur seals: A case study. *Marine Pollution Bulletin* **18**(6B), 326–335.

Fowler SL, Costa DP, Arnould JPY (2007) Ontogeny of movements and foraging ranges in the Australian sea lion. *Marine Mammal Science* **23**(3), 598–614.

Fowler SL, Costa DP, Arnould JPY, Gales NJ, Kuhn CE (2006) Ontogeny of diving behaviour in the Australian sea lion: trials of adolesence in a late bloomer. *Journal of Animal Ecology* **75**, 358–367.

Furlani D, Gales R, Pemberton D (2007) Otoliths of common temperate Australian fish: a photographic guide. (CSIRO Publishing: Melbourne)

Gales N, Cheal AJ (1992) Estimating diet composition of the Australian sea lion (*Neophoca cinerea*) from scat analysis: an unreliable technique. *Wildlife Research* **19**, 447–456.

Gales N, Hindell M, Kirkwood R (2003) 'Marine mammals: fisheries, tourism and management issues.' (CSIRO Publishing: Melbourne)

Gales N, Mattlin R (1998) Fast, safe field-portable gas anaesthesia for otariids. *Marine Mammal Science* **14**, 355–361.

Gales NJ, Cheal AJ, Pobar GJ, Williamson P (1992) Breeding biology and movements of Australian sea lions, *Neophoca cinerea*, off the west coast of Western Australia. *Wildlife Research* **19**(4), 405–416.

Gales NJ, Costa DP (1997) The Australian sea lion: an unusual life history. In 'Marine mammal research in the Southern Hemisphere: status, ecology and medicine. Vol. 1.' (Eds M Hindell and C Kemper) pp. 78–87. (Surrey Beatty & Sons: Chipping Norton, NSW)

Gales NJ, Costa DP, Kretzmann M (1996) Proximate composition of Australian sea lion milk throughout the entire supra-annual lactation period. *Australian Journal of Zoology* **44**, 651–657.

Gales NJ, Haberley B, Collins P (2000) Changes in abundance of New Zealand fur seals, *Arctocephalus forsteri*, in Western Australia. *Wildlife Research* **27**, 165–168.

Gales NJ, Shaughnessy PD, Dennis TE (1994) Distribution, abundance and breeding cycle of the Australian sea lion *Neophoca cinerea* (Mammalia: Pinnipedia). *Journal of Zoology* **234**, 353–370.

Gales R, Pemberton D (1994) Diet of the Australian fur seal in Tasmania. *Australian Journal of Marine and Freshwater Research* **45**, 653–664.

Gales R, Pemberton D, Lu CC, Clarke MR (1993) Cephalopod diet of the Australian fur seal: variation due to location, season and sample type. *Australian Journal of Marine and Freshwater Research* **44**, 657–671.

Gentry RL, Kooyman GL (1986) 'Fur seals: maternal strategies on land and at sea.' (Princeton University Press: Princeton, New Jersey)

Gibbens J, Arnould JPY (2009*a*) Age-specific growth, survival and population dynamics of female Australian fur seals. *Canadian Journal of Zoology* **87**, 902–911.

Gibbens J, Arnould JPY (2009*b*) Interannual variation in breeding chronology and pup production of Australian fur seals. *Marine Mammal Science* **25**(3), 573–587.

Gibbens J, Parry LJ, Arnould JPY (2010) Influences on fecundity in Australian fur seals (*Arctocephalus pusillus doriferus*). *Journal of Mammalogy* **91**(2), 510–518.

Gibbs CF, Tomczak M, Longmore AR (1986) The nutrient regime of Bass Strait. *Australian Journal of Marine and Freshwater Research* **37**, 451–466.

Godfrey SJ (1985) Additional observations of subaqueous locomotion in the California sea lion (*Zalophus californianus*). *Aquatic Mammals* **11**(2), 53–57.

Goldsworthy SD, Bulman C, He X, Larcombe J, Littnan CL (2003) Trophic interactions between marine mammals and Australian fisheries: an ecosystem approach. In 'Marine mammals: fisheries, tourism and management issues.' (Eds N Gales, M Hindell and R Kirkwood) pp. 62–99. (CSIRO Publishing: Melbourne)

Goldsworthy SD, McKenzie J, Page B, Lancaster ML, Shaughnessy PD, Wynen LO, Robinson SA, Peters KP, Baylis AMM, McIntosh RR (2009) Fur seals at Macquarie Island: post-sealing colonisation, trends in abundance and hybridisation of three species. *Polar Biology* **32**(10), 1473–1486.

Goldsworthy SD, Page B (2007) A risk-assessment approach to evaluating the significance of seal bycatch in two Australian fisheries. *Biological Conservation* **139**, 269–285.

Goldsworthy SD, Page B, Rogers P, Ward T (2011) 'Establishing ecosystem-based management for the South Australian Sardine Fishery: developing ecological performance indicators and reference points to assess the need for ecological allocations.' FRDC Project 2005/031 Final Report. South Australian Research and Development Institute (Aquatic Sciences), Adelaide. SARDI Publication No. F2010/000863-1. SARDI Research Report Series No. 529.

Goldsworthy SD, Page B, Shaughnessy PD, Linnane A (2010) 'Mitigating seal interactions in SRLF and gillnet sector SESSF in South Australia. Report to the Fisheries Research and Development Corporation. South Australian Research and Development Institute (Aquatic Sciences). Adelaide.

Goldsworthy SD, Pemberton D, Warneke RM (1997) Field identification of Australian and New Zealand fur seals, *Arctocephalus* spp., based in external characters. In 'Marine mammal research in the Southern Hemisphere: status, ecology and medicine. Vol. 1.' (Eds M Hindell and C Kemper) pp. 63–71. (Surrey Beatty & Sons: Chipping Norton, NSW)

Goldsworthy SD, Shaughnessy PD (1989) Subantarctic fur seals *Arctocephalus tropicalis* at Heard Island. *Polar Biology* **9**, 337–339.

Goldsworthy SD, Shaughnessy PD (1994) Breeding biology and haul-out pattern of the New Zealand fur seal, *Arctocephalus forsteri*, at Cape Gantheaume, South Australia. *Wildlife Research* **21**(3), 365–376.

Gooch R (2008) 'Seal Rocks and Victoria's primitive beginnings.' (Warrangine word: Hastings, Victoria)

Haase T (2004) 'The determinants of weaning in the New Zealand fur seal.' La Trobe University, Bundoora, Victoria

Hamer DJ, Goldsworthy SD (2006) Seal-fishery operational interactions: identifying the environmental and operational aspects of a trawl fishery that contributes to bycatch and mortality of Australian fur seals (*Arctocephalus pusillus doriferus*). *Biological Conservation* **130**, 517–529.

Hanke FD, Hanke W, Scholtyssek C, Dehnhardt G (2009) Basic mechanisms in pinniped vision. *Experimental Brain Research* **199**, 299–311.

Harcourt RG, Bradshaw CJA, Davis LS (2001) Summer foraging behaviour of a generalist predator, the New Zealand fur seal (*Arctocephalus forsteri*). *Wildlife Research* **28**(6), 599–606.

Harcourt RG, Bradshaw CJA, Dickson K, Davis LS (2002) Foraging ecology of a generalist predator, the female New Zealand fur seal. *Marine Ecology Progress Series* **227**, 11–24.

Harcourt RG, Hindell MA, Bell DG, Waas JR (2000) Three-dimensional dive profiles of free-ranging Weddell seals. *Polar Biology* **23**, 479–487.

Harris GP, Griffiths FB, Clementson LA, Lyne V, van der Doe H (1991) Seasonal and interannual variability in physical processes, nutrient cycling and the structure of the food chain in Tasmanian shelf waters. *Journal of Plankton Research* **13**, 109–131.

Harris A, Ward P (1999) 'Non-target species in Australia's commonwealth fisheries. A critical review.' Bureau of Rural Sciences: Canberra.

Harwood J, Prime JH (1978) Some factors affecting the size of British grey seal populations. *Journal of Applied Ecology* **15**, 401–411.

Hewer HR (1964) The determination of age in the grey seal (Halichoerus grypus): sexual maturity, longevity and life tables. *Proceedings of the Zoological Society of London* **142**(4), 593–623.

Higdon JW, Bininda-Emonds ORP, Beck RMD, Ferguson SH (2007) Phylogeny and divergence of the pinnipeds (Carnivora: Mammalia) assessed using a multigene dataset. *Evolutionary Biology* **7**, 216–235.

Higgins LV (1993) The nonannual, nonseasonal breeding cycle of the Australian sea lion, *Neophoca cinerea*. *Journal of Mammalogy* **74**, 270–274.

Higgins LV, Gass L (1993) Birth to weaning parturition, duration of lactation and attendance cycles of Australian sea lions, *Neophoca cinerea*. *Canadian Journal of Zoology* **71**, 2047–2055.

Hindell M, Kemper C (1997) 'Marine mammal research in the Southern Hemisphere: status, ecology and medicine.' (Surrey Beaty & Sons: Chipping Norton, NSW)

Hindell MA, Bradshaw CJA, Sumner MD, Michael KJ, Burton HR (2003) Dispersal of female southern elephant seals and their prey consumption during the austral summer: relevance to management and oceanographic zones. *Journal of Applied Ecology* **40**, 703–715.

Hindell MA, Burton HR, Slip DJ (1991) Foraging areas of southern elephant seals, *Mirounga leonina*, as inferred from water temperature data. *Australian Journal of Marine and Freshwater Research* **42**, 115–128.

Hindell MA, McConnell BJ, Fedak MA, Slip DJ, Burton HR, Reijnders PJH, McMahon CR (1999) Environmental and physiological determinants of successful foraging by naive southern elephant seal pups during their first trip to sea. *Canadian Journal of Zoology* **77**, 1807–1821.

Hindell MA, Pemberton D (1997) Successful use of a translocation program to investigate diving behavior in a male Australian fur seal, *Arctocephalus pusillus doriferus*. *Marine Mammal Science* **13**(2), 219–228.

Hume F, Arnould JPY, Kirkwood R, Davis P (2001) Extended maternal dependence by juvenile Australian fur seals (*Arctocephalus pusillus doriferus*). *Australian Mammalogy* **23**, 67–70.

Hume F, Hindell MA, Pemberton D, Gales R (2004) Spatial and temporal variation in the diet of a high trophic level predator, the Australian fur seal (*Arctocephalus pusillus doriferus*). *Marine Biology* **144**, 407–415.

Hume F, Pemberton D, Gales R, Brothers N, Greenwood M (2002) Trapping and relocating seals from salmonid fish farms in Tasmania, 1990–2000: was it a success? *Papers and Proceedings of the Royal Society of Tasmania* **136**, 1–6.

Hyvärinen H (1989) Diving in darkness: Whiskers as sense organs of the ringed seal (*Phoca hispida saimensis*). *Journal of the Zoological Society, London* **218**, 663–678.

Jones FW (1925) 'The mammals of South Australia Part 2: The Monodelphia.' (South Australian Government Printer: Adelaide)

Kailola PJ, Williams MJ, Stewart PC, Reichelt RE, McNee A, Grieve C (1993) 'Australian fisheries resources.' (Bureau of Rural Science and Fisheries Resource Development Corporation Publication: Canberra)

Kämpf J, Doubell M, Griffin D, Matthews RL, Ward TM (2004) Evidence of a large seasonal coastal upwelling system along the southern shelf of Australia. *Geophysical Research Letters* **31**, 1–4.

Kastelein RA, Stevens S, Mostard P (1990) The tactile sensitivity of the mystacial vibrissae of a Pacific walrus (*Odobenus rosmarus divergens*). Part 2: Masking. *Aquatic Mammals* **16**(2), 78–87.

Kemper C, Pemberton D, Cawthorn M, Heinrich S, Wursig B, Shaughnessy P, Gales R (2003) Aquaculture and marine mammals: co-existence or conflict? In 'Marine mammals: fisheries, tourism and management issues.' (Eds N Gales, M Hindell and R Kirkwood) pp. 208–225. (CSIRO Publishing: Melbourne)

King JE (1969) The identity of seals in Australia. *Australian Journal of Zoology* **17**(5), 841–853.

King JE (1983) 'Seals of the world: 2nd edition.' (Oxford University Press: London)

Kirkman SP, Oosthuizen WH, Meyer MA, Kotze PGH, Roux JP, Underhill LG (2007) Making sense of censuses and dealing with missing data: trends in pup counts of Cape fur seal *Arctocephalus pusillus pusillus* for the period 1972–2004. *African Journal of Marine Science* **29**(2), 161–176.

Kirkwood R, Arnould JPY (2011) Foraging trip strategies and habitat use during late pup rearing by lactating Australian fur seals. *Australian Journal of Zoology* **59**, 216–226.

Kirkwood R, Boren L, Shaughnessy P, Szteren D, Mawson P, Hückstädt L, Hofmeyr G, Oosthuizen H, Campagna C, Berris M (2003) Pinniped-focused tourism in the Southern Hemisphere: a review of the industry. In 'Marine mammals and humans: fisheries, tourism and management issues.' (Eds N Gales, M Hindell and R Kirkwood) pp. 257–276. (CSIRO Publishing: Melbourne)

Kirkwood R, Dickie J (2005) Mobbing of a great white shark (Carcharodon carcharias) by adult male Australian fur seals (Arctocephalus pusillus doriferus). *Marine Mammal Science* **21**(2), 336–339.

Kirkwood R, Gales N, Lynch M, Dann P (2002) Satellite tracker deployments on adult, male Australian fur seals *Arctocephalus pusillus doriferus*: methods and preliminary results. *Australian Mammalogy* **24**, 73–83.

Kirkwood R, Gales R, Terauds A, Arnould JPY, Pemberton D, Shaughnessy PD, Mitchell AT, Gibbens J (2005) Pup production and population trends of the Australian fur seal *Arctocephalus pusillus doriferus*. *Marine Mammal Science* **21**, 260–282.

Kirkwood R, Hume F, Hindell M (2008) Sea temperature variations mediate annual changes in the diet of Australian fur seals in Bass Strait. *Marine Ecology Progress Series* **369**, 297–309.

Kirkwood R, Lynch M, Gales N, Dann P, Sumner M (2006) Foraging strategies of an adult, male otariid: the Australian fur seal (*Arctocephalus pusillus doriferus*). *Canadian Journal of Zoology* **84**(12), 1781–1788.

Kirkwood R, Pemberton D, Gales R, Hoskins AJ, Mitchell A, Shaughnessy PD, Arnould JPY (2010) Continued population recovery by Australian fur seals. *Marine and Freshwater Research* **61**, 695–701.

Kirkwood R, Warneke RM, Arnould JPY (2009) Recolonization of Bass Strait, Australia, by the New Zealand fur seal *Arctocephalus forsteri*. *Marine Mammal Science* **25**(2), 441–449.

Knuckey IA, Eayrs S, Bosschietter B (2002) Options for reducing incidental catch of seals on wet-boats in the SETF: a preliminary assessment. Final report to the Australian Fisheries Management Authority. Marine and Freshwater Resources Institute, Queenscliff, Victoria.

Kooyman GL (1973) Respiratory adaptations in marine mammals. *American Zoologist* **13**, 457–468.

Kooyman GL (1989) 'Diverse divers: physiology and behaviour.' (Springer-Verlag: New York)

Kretzmann MB, Costa DP, Higgins LV, Needham DJ (1991) Milk composition of Australian sea lions *Neophoca Cinerea*, variability in lipid content. *Canadian Journal of Zoology* **69**, 2556–2561.

Lake S (1997) Analysis of the diet of New Zealand fur seals *Arctocephalus forsteri* in Tasmania. In 'Marine mammal research in the Southern Hemisphere: status, ecology and medicine. Vol. 1.' (Eds M Hindell and C Kemper) pp. 125–129. (Surrey Beatty & Sons: Chipping Norton, NSW)

Lalas C, Bradshaw CJA (2001) Folklore and chimerical numbers: review of a millennium of interaction between fur seals and humans in the New Zealand region. *New Zealand Journal of Marine and Freshwater Research* **35**, 477–497.

Lancaster ML, Arnould JPY, Kirkwood R (2010) Genetic status of an endemic marine mammal, the Australian fur seal, following historical harvesting. *Animal Conservation* **13**(3), 247–255.

Lancaster ML, Gemmell NJ, Negro S, Goldsworthy S, Sunnucks P (2006) Ménage à trois on Macquarie Island: hybridization among three species of fur seal (*Arctocephalus* spp.) following historical population extinction. *Molecular Ecology* **15**, 3681–3692.

Lavigne DM, Scheffer VB, Kellert SR (1999) The changing place of marine mammals in American thought. In 'Conservation and managment of marine mammals.' (Eds JR Twiss and RR Reeves). (Melbourne University Press: Melbourne)

Le Souef AS (1925) Notes on the seals found in Australian seas. *Australian Zoologist* **4**, 112–116.

Lea M-A, Hindell MA (1997) Pup growth and maternal care in New Zealand fur seals, *Arctocephalus forsteri*, at Maatsuyker Island, Tasmania. *Wildlife Research* **24**, 307–318.

Lento GM, Haddon M, Chambers GK, Baker CS (1997) Genetic variation of Southern Hemisphere fur seals (*Arctocephalus* spp.): investigation of population structure and species identity. *The Journal of Heredity* **88**, 202–208.

Lento GM, Mattlin RH, Chambers GK, Baker CS (1994) Geographic distribution of mitochondrial cytochrome b DNA haplotypes in New Zealand fur seals (*Arctocephalus forsteri*). *Canadian Journal of Zoology* **72**, 293–299.

Lewis F (1930) Seals on the Victorian coast and their feeding habits. *Australian Museum Magazine* **4**, 39–44.

Ling JK (1965) Functional significance of sweat glands and sebaceous glands in seals. *Nature* **208**, 560–562.

Ling JK (1970) Pelage and molting in wild animals with special reference to aquatic forms. *The Quarterly Review of Biology* **45**, 16–54.

Ling JK (1977) Vibrissae of marine mammals. In 'Functional anatomy of marine mammals.' Ed. RJ Harrison). (Academic Press: London)

Ling JK (1999) Exploitation of fur seals and sea lions from Australia, New Zealand and adjacent subantarctic islands during the eighteenth, nineteenth and twentieth centuries. *Australian Zoologist* **31**(2), 323–350.

Ling JK (2002) Impact of colonial sealing on seal stocks around Australia, New Zealand and subantarctic islands between 150 and 170 degrees east. *Australian Mammalogy* **24**, 117–126.

Ling JK, Walker GE (1978) An 18-month breeding cycle in the Australian sea lion? *Search* **9**, 464–465.

Littnan CL (2003) Approaches to studying the foraging ecology of the Australian fur seal *Arctocephalus pusillus doriferus* in northern Bass Strait. PhD thesis, Macquarie University, Sydney

Littnan CL, Arnould JPY (2002) At-sea movements of female Australian fur seals (*Arctocephalus pusillus doriferus*). *Australian Mammalogy* **24**, 65–72.

Littnan CL, Arnould JPY, Harcourt RG (2007) Effect of proximity to the shelf edge on the diet of female Australian fur seals. *Marine Ecology Progress Series* **338**, 257–267.

Littnan CL, Mitchell AT (2002) Australian and New Zealand fur seals at the Skerries, Victoria: recovery of a breeding colony. *Australian Mammalogy* **24**, 57–64.

Lovasc T, Croft DB, Banks P (2008) Establishing tourism guidelines for viewing Australian sea lions Neophoca cinerea at Seal Bay Conservation Park, South Australia. In 'Too close for comfort: contentious issues in human-wildlife encounters.' (Eds D Lunney, A Munn and W Meikle) pp. 225–232. (Royal Zoological Society of New South Wales: Mosman, NSW)

Lowther AD, Goldsworthy SD (2011) Maternal strategies of the Australian sea lion (*Neophoca cinerea*) at Dangerous Reef, South Australia. *Australian Journal of Zoology* **59**(1), 54–62.

Lowther AD, Harcourt RG, Goldsworthy SD, Stow A (2012) Population structure of adult female Australian sea lions is driven by fine-scale foraging site fidelity. *Animal Behaviour* **83**, 691–701.

Lyle JM, Willcox ST (2008) Dolphin and seal interactions with mid-water trawling in the Commonwealth small pelagic fishery, including an assessment of bycatch mitigation. Final Report Project R05/0996 to AFMA, Tasmanian Aquaculture and Fisheries Institute, University of Tasmania, Hobart.

Lynch M, Duignan PJ, Taylor T, Nielsen O, Kirkwood R, Gibbens J, Arnould JPY (2011) Epidemiology of *Brucella* infection in Australian fur seals. *Journal of Wildlife Diseases* **47**(2), 352–363.

Lynch M, Kirkwood R, Gray R, Robson D, Burton G, Jones L, Sinclair R, Arnould JPY (2012) Characterization and causality investigations of an alopecia syndrome in Australian fur seals (*Arctocephalus pusillus doriferus*). *Journal of Mammalogy* **93**(2), 504–513.

Lynch M, Kirkwood R, Mitchell A, Arnould JPY (2011) Prevalence and significance of an alopecia syndrome in Australian fur seals (*Arctocephalus pusillus doriferus*). *Journal of Mammalogy* **92**(2), 342–351.

Lynch M, Nielsen O, Duignan PJ, Kirkwood R, Hoskins AJ, Arnould JPY (2011) Serological survey for potential pathogens and assessment of disease risk in Australian fur seals. *Journal of Wildlife Diseases* **47**(3), 555–565.

Mawson PR, Coughran DK (1999) Records of sick, injured and dead pinnipeds in Western Australia 1980–1996. *Journal of the Royal Society of Western Australia* **82**, 121–128.

McCoy F (1885) 'Prodromus of the zoology of Victoria.' (J. Ferres, Government Printer: Melbourne)

McIntosh RR (2007a) 'The life history and population demographics of the Australian sea lion, *Neophoca cinerea*.' (La Trobe University: Bundoora, Victoria)

McIntosh RR (2007b) Louse infestations of the Australian sea lion *Neophoca cinerea*. *Australian Mammalogy* **29**, 103–106.

McIntosh R, Goldsworthy SD, Shaughnessy PD, Kennedy CW, Burch P (2012) Estimating pup production in a mammal with an extended and aseasonal breeding season, the Australian sea lion (*Neophoca cinerea*). *Wildlife Research* **39**, 137–148.

McIntosh RR, Page B, Goldsworthy SD (2006) Dietary analysis of regurgitates and stomach samples from free-living Australian sea lions. *Wildlife Research* **33**, 661–669.

McKenzie J, Page B, Goldsworthy SD, Hindell MA (2007) Growth strategies of New Zealand fur seals (*Arctocephalus forsteri*). *Journal of Zoology* **272**, 377–389.

McKenzie J, Parry LJ, Page B, Goldsworthy SD (2005) Estimation of pregnancy rates and reproductive failure in New Zealand fur seals (*Arctocephalus forsteri*). *Journal of Mammalogy* **86**, 1237–1246.

Menkhorst PM (1995) 'Mammals of Victoria: distribution, ecology and conservation.' (Oxford University Press: Melbourne)

Middleton JF, Arthur C, van Ruth P, Ward TM, McClean JL, Maltrud ME, Gill P, Levings A, Middleton S (2007) El Niño effects and upwelling off South Australia. *Journal of Physical Oceanography* **37**, 2458–2477.

Middleton JF, Bye JAT (2007) A review of shelf-slope circulation along Australia's southern shelves: Cape Leeuwin to Portland. *Progress in Oceanography* **75**, 1–41.

Middleton JF, Cirano M (2002) A northern boundary current along Australia's southern shelves: the Flinders Current. *Journal of Geophysical Research* **107**(C9), 3129–3140.

Muir SF, Barnes DKA, Reid K (2006) Interactions between humans and leopard seals. *Antarctic Science* **19**(1), 61–74.

Mukhametov LM (1985) Unihemispheric slow wave sleep in the brain of dolphins and seals. In 'Endogenous sleep substances and sleep regulation.' (Eds S Inoue and AA Borbely) pp. 67–75. (YNU Science Press: Utrecht)

Needham DJ (1997) The role of stones in the sea lion stomach: investigations using contrast radiography and fluoroscopy. In 'Marine mammal research in the Southern Hemisphere. Vol. 1.' (Eds M Hindell and C Kemper). (Surrey Beatty & Sons: Chipping Norton, NSW)

Orsini J-P (2004) Human impacts on Australian sea lions, Neophoca cinerea, hauled out on Carnac Island (Perth, Western Australia): implications for wildlife and tourism management. (Murdoch University: Perth)

Pabst DA, Rommel SA, McLellan WA (1999) The functional morphology of marine mammals. In 'Biology of marine mammals.' (Eds JE Reynolds and SA Rommel). (Melbourne University Press: Melbourne)

Page B, Goldsworthy SD, Hindell M (2002) Vocal traits of mother and pup fur seals (*Arctocephalus* spp.). *Bioacoustics* **13**, 121–143.

Page B, McKenzie J, *et al.* (2004) Entanglement of Australian sea lions and New Zealand fur seals in lost fishing gear and other marine debris before and after Government and industry attempts to reduce the problem. *Marine Pollution Bulletin* **49**, 33–42.

Page B, McKenzie J, Goldsworthy SD (2005*a*) Dietary resource partitioning among sympatric New Zealand and Australian fur seals. *Marine Ecology Progress Series* **293**, 283–302.

Page B, McKenzie J, Goldsworthy SD (2005*b*) Inter-sexual differences in New Zealand fur seal diving behaviour. *Marine Ecology Progress Series* **304**, 249–264.

Page B, McKenzie J, Hindell MA, Goldsworthy SD (2005) Drift dives by male New Zealand fur seals (*Arctocephalus forsteri*). *Canadian Journal of Zoology* **83**, 293–300.

Page B, McKenzie J, Summer MD, Coyne M, Goldsworthy S (2006) Spatial separation of foraging habitats among New Zealand fur seals. *Marine Ecology Progress Series* **323**, 263–279.

Page B, Welling A, Chambellant M, Goldsworthy SD, Dorr T, van Veen R (2003) Population status and breeding season chronology of Heard Island fur seals. *Polar Biology* **26**(4), 219–224.

Pearse RJ (1979) Distribution and conservation of the Australian fur seal in Tasmania. *Victorian Naturalist* **96**, 48–53.

Pemberton D, Brothers NP, Kirkwood R (1992) Entanglement of Australian fur seals in man-made debris in Tasmanian waters. *Wildlife Research* **19**, 151–159.

Pemberton D, Gales R (2004) Australian fur seals (*Arctocephalus pusillus doriferus*) breeding in Tasmania: population size and status. *Wildlife Research* **31**, 301–309.

Pemberton D, Kirkwood RJ (1994) Pup production and distribution of the Australian fur seal, *Arctocephalus pusillus doriferus*, in Tasmania. *Wildlife Research* **21**, 341–352.

Pemberton D, Shaughnessy PD (1993) Interaction between seals and marine fish-farms in Tasmania, and management of the problem. *Aquatic Conservation* **3**, 149–158.

Pemberton D, Kirkwood R, Gales R, Renouf D (1993) Size and shape of male Australian fur seals, *Arctocephalus pusillus doriferus*. *Marine Mammal Science* **9**(1), 99–103.

Pitcher BJ, Harcourt RG, Schaal B, Charrier I (2011) Social olfaction in marine mammals: wild female Australian sea lions can identify their pup's scent. *Biology Letters* **7**, 60–62.

Poloczanska ES, Babcock RC, *et al.* (2007) Climate change and Australian marine life. *Oceanography and Marine Biology: An Annual Review* **45**, 407–478.

Prince JD (2001) Ecosystem of the South East Fishery (Australia), and fisher lore. *Marine and Freshwater Research* **52**, 431–449.

Punt AE, David JHM, Leslie RW (1995) The effects of future consumption by the Cape fur seal on catches and catch rates of the Cape hakes. 2. Feeding and diet of the

Cape fur seal *Arctocephalus pusillus pusillus*. *South African Journal of Marine Science* **16**, 85–99.

Raga JA, Balbuena JA, Aznar FJ, Fernandez M (1997) The impact of parasites on marine mammals: a review. *Parassitologia* **39**, 293–296.

Rawson G (1946) 'Matthew Flinders narrative of his voyage in the schooner Francis: 1798, preceded and followed by notes on Flinders, Bass, the wreck of the Sydney Cove etc.' (Golden Cockerel Press: London) p. 100

Renwick L, Kirkwood R (2004) An extended visit by a leopard seal to Phillip Island, Victoria. *Victorian Naturalist* **121**, 55–59.

Repenning CIRCA, Peterson RS, Hubbs CL (1971) Contributions to the systematics of the southern fur seals with particular reference to the Juan Fernandez and Guadelupe species. In 'Antarctic pinnipedia.'. (Ed. WE Burt) pp. 1–52. (American Geophysical Union: Washington, DC)

Richards R (1994) The "upland seal" of the Antipodes and Macquarie Islands: a historian's perspective. *Journal of the Royal Society of New Zealand* **24**(3), 289–295.

Riedman M (1990) 'The pinnipeds: seals, sea lions and walruses.' (University of California Press: Berkley)

Robertson BC, Gemmell NJ (2005) Microsatellite DNA markers for the study of population structure in the New Zealand fur seal *Arctocephalus forsteri*. New Zealand Department of Conservation Science Internal Series 196, Wellington.

Robinson S, Gales R, Terauds A, Greenwood M (2008) Movements of fur seals following relocation from fish farms. *Aquatic Conservation: Marine and Freshwater Ecosystems* **18**, 1189–1199.

Robinson S, Terauds A, Gales R, Greenwood M (2008) Mitigating fur seal interactions: relocation from Tasmanian aquaculture farms. *Aquatic Conservation: Marine and Freshwater Ecosystems* **18**, 1180–1188.

Rogers TL (2003) Factors influencing the acoustic behaviour of male phocid seals. *Aquatic Mammals* **29**(2), 247–260.

Rounsevell D, Pemberton D (1994) The status and seasonal occurence of leopard seals, *hydrurga leptonyx*, in Tasmanian waters. *Australian Mammalogy* **17**, 97–102.

Roux JP (1987) Recolonization processes in the subantarctic fur seal, *Arctocephalus tropicalis*, on Amsterdam Island. *NOAA Technical Report NMFS* **51**, 189–194.

Schahinger R (1987) Structure of coastal upwelling events observed off the southeast coast of South Australia. *Australian Journal of Marine and Freshwater Research* **38**, 439–459.

Shaughnessy PD (1970) Serum protein variation in southern fur seals, *Arctocephalus* spp., in relation to their taxonomy. *Australian Journal of Zoology* **18**, 331–343.

Shaughnessy PD (1999) The action plan for Australian seals. Environment Australia, Biodiversity Group, Threatened Species and Communities Section, Canberra.

Shaughnessy PD, Davenport SR (1996) Underwater videographic observations and incidental mortality of fur seals around fishing equipment in South-eastern Australia. *Australian Journal of Marine and Freshwater Research* **47**, 553–556.

Shaughnessy PD, Gales NJ, Dennis TE, Goldsworthy SD (1994) Distribution and abundance of New Zealand fur seals, *Arctocephalus forsteri*, in South Australia and Western Australia. *Wildlife Research* **21**, 667–695.

Shaughnessy PD, Goldsworthy SD (1990) Population size and breeding season of the Antarctic fur seal *Arctocephalus gazella* at Heard Island – 1987/88. *Marine Mammal Science* **6**(4), 292–304.

Shaughnessy PD, Goldsworthy SD, Hamer DJ, Page B, McIntosh RR (2011) Australian sea lions *Neophoca cinerea* at colonies in South Australia: distribution and abundance, 2004 to 2008. *Endangered Species Research* **13**, 87–98.

Shaughnessy PD, Goldsworthy SD, Libke JA (1995) Changes in the abundance of New Zealand fur seals, *Arctocephalus forsteri*, on Kangaroo Island, South Australia. *Wildlife Research* **22**, 201–215.

Shaughnessy P, Kirkwood R, Cawthorn M, Kemper C, Pemberton D (2003a) Pinnipeds, cetaceans and fisheries in Australia: a review of operational interactions. In 'Marine mammals: fisheries, tourism and management issues.' (Eds N Gales, M Hindell and R Kirkwood) pp. 136–152. (CSIRO Publishing: Melbourne)

Shaughnessy P, Kirkwood R, Pemberton D, Cawthorn M (2003b) Pinnipeds, cetaceans and Australian fisheries: a review of operational interactions. In 'Marine mammals and humans: fisheries, tourism and management issues.' (Eds N Gales, M Hindell and R Kirkwood) pp. 136–152. (CSIRO Publishing: Melbourne)

Shaughnessy PD, Kirkwood RJ, Warneke RM (2002) Australian fur seals, *Arctocephalus pusillus doriferus*: pup numbers at Lady Julia Percy Island, Victoria, and a synthesis of the species' population status. *Wildlife Research* **29**(2), 185–192.

Shaughnessy PD, McKenzie J, Lancaster ML, Goldsworthy SD, Dennis TE (2010) Australian fur seals establish haulout sites and a breeding colony in South Australia. *Australian Journal of Zoology* **58**, 94–103.

Shaughnessy PD, McKeown A (2002) Trends in abundance of New Zealand fur seals, *Arctocephalus forsteri*, at the Neptune Islands, South Australia. *Wildlife Research* **29**, 363–370.

Shaughnessy PD, Shaughnessy GL, Fletcher L (1988) Recovery of the fur seal population at Macquarie Island. *Papers and Proceedings of the Royal Society of Tasmania* **122**(1), 177–187.

Shaughnessy PD, Testa JW, Warneke RM (1995) Abundance of Australian fur seal pups, *Arctocephalus doriferus*, at Seal Rocks, Victoria, in 1991–92 from Petersen and Bayesian estimators. *Wildlife Research* **22**, 625–632.

Shaughnessy PD, Troy SK, Kirkwood R, Nicholls AO (2000) Australian fur seals at Seal Rocks, Victoria: pup abundance by mark-recapture estimation shows continued increase. *Wildlife Research* **27**, 629–633.

Southwell CJ, Paxton CGM, Borchers DL, Boveng PL, Nordøy ES, Blix AS, De La Mare WK (2008) Estimating population status under conditions of uncertainty: the Ross seal in East Antarctica. *Antarctic Science* **20**(2), 123–133.

Spence-Bailey LM, Verrier D, Arnould JPY (2007) The physiological and behavioural development of diving in Australian fur seal (*Arctocephalus pusillus doriferus*) pups. *Journal of Comparative Physiology B* **177**, 483–494.

Staniland IJ (2002) Investigating the biases in the use of hard prey remains to identify diet composition using Antarctic fur seals (*Arctocephalus gazella*) in captive feeding trials. *Marine Mammal Science* **22**, 223–243.

Stewardson CL, Bester MN, Oosthuizen WH (1998) Reproduction in the male Cape fur seal *Arctocephalus pusillus pusillus*: age at puberty and annual cycle of the testis. *Journal of Zoology* **246**, 63–74.

Stirling I, Warneke RM (1971) Implications of a comparison of the airborne vocalizations and some aspects of the behaviour of the two Australian fur seals, *Arctocephalus* spp., on the evolution and present taxonomy of the genus. *Australian Journal of Zoology* **19**, 227–241.

Stockton J (1982) Seals in Tasmanian prehistory. *Proceedings of the Royal Society of Victoria* **94**, 53–60.

Taylor RH (1982) New Zealand fur seals at the Bounty Islands. *New Zealand Journal of Marine and Freshwater Research* (16), 1–9.

Thresher RE (1994) Climatic cycles may explain fish recruitment in south east Australia. *Australian Fisheries* **53**, 20–22.

Tilzey RDJ, Rowling KR (2001) History of Australia's south east fishery: a scientist's perspective. *Marine and Freshwater Research* **52**, 361–375.

Tripovich JS, Charrier I, Rogers TL, Canfield R, Arnould JPY (2008) Who goes there? The dear-enemy effect in male Australian fur seals (*Arctocephalus pusillus doriferus*). *Marine Mammal Science* **24**(4), 941–948.

Tripovich JS, Rogers TL, Arnould JPY (2005) Species-specific characteristics and individual variation of the bark call produced by male Australian fur seals, *Arctocephalus pusillus doriferus*. *Bioacoustics* **15**(1), 79–96.

Tripovich JS, Rogers TL, Canfield R, Arnould JPY (2006) Individual variation in the pup attraction call produced by female Australian fur seals during early lactation. *The Journal of the Acoustical Society of America* **120**(1), 79–96.

Troy SK, Mattlin R, Shaughnessy PD, Davie PS (1999) Morphology, age and survival of adult male New Zealand fur seals, *Arctocephalus forsteri*, in South Australia. *Wildlife Research* **26**(1), 21–34.

van Dyck S, Strahan R (2008) 'The mammals of Australia: 3rd edition.' (Reed New Holland: Sydney)

Ward TM, Hoedt F, McLeay L, Dimmlich WF, Kinloch M, Jackson GD, McGarvey R, Rogers PJ, Jones K (2001) Effects of the 1995 and 1998 mass mortality events on the spawning biomass of sardine, *Sardinops sagax*, in South Australian waters. *ICES Journal of Marine Science* **58**, 865–875.

Ward TM, McLeay LJ, Dimmlich WF, Rogers PJ, McClatchie S, Matthews R, Kämpf J, van Ruth PD (2006) Pelagic ecology of a northern boundary current system: effects of upwelling on the production and distribution of sardine (*Sardinops sagax*), anchovy (*Engraulis australis*) and southern bluefin tuna (*Thunnus maccoyii*) in the Great Australian Bight. *Fisheries Oceanography* **15**(3), 191–207.

Warneke RM (1975) Dispersal and mortality of juvenile fur seals *Arctocephalus pusillus doriferus* in Bass Strait, southeastern Australia. *Rapports et Proces-Verbaux des Reunions - Conseil International pour L'Exploration de la Mer* **169**, 296–302.

Warneke RM (1982) The distribution and abundance of seals in the Australasian region, with summaries of biology and current research. In 'Mammals in the seas, FAO fisheries series No.5. Vol. IV.' pp. 431–474. (FAO: Rome)

Warneke RM (2002) Seals at Seal Rocks, Western Port, and in Bass Strait, before and after the Baudin expedition's visit in 1802. In '*Le Naturaliste* in Western Port 1802–2002.'. (Eds N Macwhirter, P Macwhirter, JL Sagliocco and J Southwood) pp. 77–98. (Department of Infrastructure: Melbourne)

Warneke RM, Shaughnessy PD (1985) *Arctocephalus pusillus*, the South African and Australian fur seal: taxonomy, evolution, biogeography, and life history. In 'Studies of sea mammals in south latitudes.' (Eds JK Ling and MM Bryden) pp. 53–77. (South Australian Museum: Adelaide)

Wartzok D, Ketten DR (1999) Marine mammal sensory systems. In 'Biology of marine mammals.' (Eds JE Reynolds and SA Rommel). (Melbourne University Press: Melbourne)

Watkins WA, Wartzok D (1985) Sensory biophysics of marine mammals. *Marine Mammal Science* **1**(3), 219–260.

White WB, Peterson RG (1996) An Antarctic circumpolar wave in surface pressure, wind, temperature and sea-ice extent. *Nature* **380**, 699–702.

Wickens P, York AE (1997) Comparative population dynamics of fur seals. *Marine Mammal Science* **13**(2), 241–292.

Wilson RP, McMahon CR (2006) Measuring devices on wild animals: what constitutes acceptable practice? *The Ecological Society of America* **4**(3), 147–154.

Woods R, Cousins DV, Kirkwood R, Obendorf DL (1995) Tuberculosis in a wild Australian fur seal (*Arctocephalus pusillus doriferus*) from Tasmania. *Journal of Wildlife Diseases* **31**(1), 83–86.

Wynen LP, Goldsworthy SD, *et al.* (2001) Phylogenetic relationships within the eared seals (Otariidae: Carnivora): Implications for the historical biogeography of the family. *Molecular Phylogenetics and Evolution* **21**, 270–284.

Yodzis P (2001) Must top predators be culled for the sake of fisheries? *Trends in Ecology & Evolution* **16**(2), 78–84.

Yonezawa T, Kohno N, Hasegawa M (2009) The monophyletic origin of sea lions and fur seals (Carnivora; Otariidae) in the Southern Hemisphere. *Gene* **441**, 89–99.

INDEX

www.ingramcontent.com/pod-product-compliance
Lightning Source LLC
Chambersburg PA
CBHW041130280526
45792CB00013B/2369